I0074574

The essential

TOILET BOOK

of

PHYSICS

G.A. LEWIS

www.essentialtoiletbooks.com

First published in Great Britain in 2021

Copyright © G.A. Lewis

The moral right of this author has been asserted.

All rights reserved.

No part of this publication may be reproduced, stored in a retrieval system, or transmitted, in any form or by any means, without the prior permission in writing of the publisher, nor be otherwise circulated in any form of binding or cover other than that in which it is published and without a similar condition including this condition being imposed on the subsequent purchaser.

Typesetting and publishing by UK Book Publishing

www.ukbookpublishing.com

ISBN: 978-1-914195-33-4

To my parents Dave and Marriee .... for
my education, and everything else.

"The mind is not a vessel to be filled, but a fire to be lighted"
Plutarch

It is my deepest hope that this small, run-of-the-mill
book will one day be picked up by some inquisitive
youngster, who having found it lying around,
will be sufficiently intrigued and unknowingly
inspired as to one day go on to make some great
or humble contribution to the benefit of all.

Contents

Foreword

"With science ..... one of the things that makes it very difficult is that it takes a lot of imagination. It's very hard to imagine all the crazy things that things really are like": Richard Feynman[1]

Two guys are in a pub. Over a few beers, one of them is explaining that as far as he is concerned, the world is flat, suspended in the universe on the back of a giant turtle. With patience his friend explains that actually the world is round, like a football, with most of the people either sideways or upside down, spinning like a top whilst at the same time hurtling around the sun at 100,000 km per hour. Not only that, the whole complete solar system of the Sun, the planets, and the spinning Earth with its upside down people is ploughing through space faster still, on a gigantic orbit around the galaxy so big it takes 230 million years to go around just once. Quietly his friend takes a sip of his beer. He's not convinced. He prefers the giant turtle.

It is often said that reality is stranger than fiction, and nowhere else is this truer than in the world of physics. It is a world where many of even the most basic concepts can escape the imagination, and where the most attractive details can get lost in a cloud of advanced mathematics impenetrable to all but the most dedicated. However underneath all the formal theories lies an incredible and fascinating story, a story that should be as familiar to us as any children's fairy tale, or history of kings, queens and empires.

The story of physics deserves to be told not only because it is fascinating, but because it is relevant. As we advance through the first part of the 21st century the pace of the technological revolution only quickens, and basic physics is at the very heart of the most important and significant of the emerging technologies destined to change our world, from quantum computing to artificial intelligence, nanotechnologies to virtual reality, superconductivity to fusion power.[2]

This small book sets out to tell a little of this incredible story, and to familiarize with some of the most basic concepts that crop up constantly, in the media, in the movies, or in our kids' homework projects. It supposes no prior knowledge nor background in the subject, only a genuine curiosity, disposition, and an open mind. Enjoy!

The Theory of Everything

*"My goal is simple. It is a complete understanding of the universe,
why it is as it is and why it exists at all": Stephen Hawking*

As many may remember, the title of this first chapter was also the title of the excellent 2014 film on the life of Stephen Hawking, superbly portrayed by Eddie Redmayne, earning him a well-deserved Oscar in the process. A central theme of the film was Hawking's doggedness to arrive at the *theory of everything,* a complete theory that could be fully expressed in one simple equation, containing absolutely everything we know about the workings of the universe and all within in it.

Of course Hawking was not alone in this, since the search for such an equation has been the Holy Grail of physicists for decades. However it may come as a surprise to know that actually, this equation exists! So brace yourself dear reader, for here it comes:[1]

Theory of Everything (so far!!)

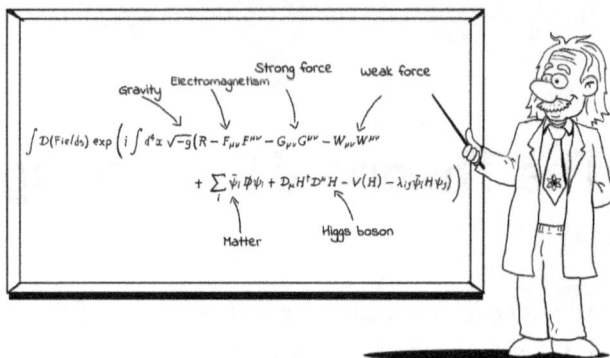

Well, it qualifies as being one equation, although simple it certainly is not. Neither does it explains absolutely everything. There are certain phenomena in our universe where this marvellous equation falls a little short, such as accounting for dark matter or dark energy, or providing a complete description of black holes, to name but a few. Neither of course does it explain natural selection and the evolution of the species, nor the pros and cons of different economic models, nor how to write a successful opera or how best to bring up our kids. These are ideas and concepts emergent on the larger macroscopic human level.[2] What this equation attempts to incorporate is everything we know and understand on the most basic microscopic level, and to be a compendium of all our insights on the most fundamental physical processes of the universe that surrounds us.

Despite its shortcomings, marvellous it certainly is. This one single equation contains the formulas for understanding absolutely everything in our everyday lives, all we can see, hear, touch, smell and taste. It correctly predicts the outcome of every single scientific experiment ever made, and no investigation has ever been carried out where this equation has failed to work. It has never been proved wrong, and sums up all we currently know, from the microcosm of sub atomic particles, to the macrocosmic structure of the universe. The equation is the very tip of a mathematical iceberg, containing within it all the success stories developed and built up since the beginnings of modern science.[1]

The part relating to gravity started with Isaac Newton and his *Principia* of 1687, which was subsequently submitted to a significant rewrite in the early 1900's by the young Albert Einstein. Modern electromagnetism kicked off in the early 1800's based on the work and investigations of scientists like Michael Faraday, and pulled together in 1873 by James Clarke Maxwell and his landmark *Treatise on Electricity and Magnetism*. Their ideas subsequently fed the quantum revolution of the 20th century, with each generation of physicists refining and deepening our understandings. So what we see when we look at this equation is really a collection of all the ideas, research, experiments and discoveries of the whole field of physics throughout the centuries, resulting in this wonderful snapshot of where we are today.

Often referred to as the Core Theory, this eclectic equation is built on the two pillars of modern physics: Einstein's general theory of relativity, and quantum mechanics, or 'quantum theory'. Whilst general relativity was the work of one single individual, quantum mechanics is the result of contributions from some of the 20th century's greatest physicists, such as Bohr, Heisenberg, Schrodinger, Dirac, Feynman, Gell-Mann and many others. Even Einstein was involved, but more on that later.

CORE THEORY = GENERAL RELATIVITY + QUANTUM MECHANICS

General relativity is our best theory on the workings of gravity, and can predict and explain happenings on the very largest of scales, across great distances, involving objects of significant mass. The classical theory of Newton could tell us *how* gravity worked, so predicting which way the apple would fall, or calculating that it takes the Earth 365 and a bit days to orbit the Sun. Einstein's relativity went one step further, giving us a vision as to *why* gravity works the way it does, and in the process shedding light on much wider phenomena helping us to a more complete understanding of our universe. It explains how light traveling in space can be bent, tells us why identical clocks can run at different rates, helps us understand the formation of stars, galaxies, black holes, and indeed established the whole framework of modern cosmology and astrophysics.

Quantum mechanics on the other hand underpins our understanding of the extremely small. It tells us about the sub atomic particles that everything is made of, how they interact, stick together, blow apart, and give us phenomena such as electricity, magnetism and nuclear energy. Quantum has been the theoretical basis for our development of integrated circuits, computers, and indeed the whole field of electronics and communication technologies that has so revolutionized our world.

Despite all its successes, physicists continue to have issues with certain parts of the equation, in particular the inclusion of gravity and its relation to the other fundamental forces which has yet to be fully ironed out. Another objection comes from its 'awkwardness', from the sense that it has somehow been cobbled together from all the underlying theories which it contains, and far from the simplicity and elegance subjectively associated with any truly fundamental expression of reality. Consequently the work goes on in the world of physics, to fully come to grips with the equation, and to try to improve it, to simplify it, to beautify it, or even to come up with a completely new and more fundamental set of ideas with which to replace it. However for now it is the very best thing we have, the very zenith of our scientific knowledge, and the current pinnacle of the reductionist approach to science so valued by Hawking and his contemporaries.

1905 'Annus Mirabilis'

"Imagination is more important than knowledge": Albert Einstein

In 1905, Albert Einstein, then aged 26, was working as a third grade clerk at the Swiss Patent Office in Bern. He'd recently applied for a second grade position, but his application had been rejected. His job did afford him however a certain amount of 'hobby' time, and in that one year he submitted five papers to the prestigious German scientific journal *Annalen der Physik*. Two papers were based on work carried out with fluids, providing the first accurate measurements of the dimensions of molecules, and incredibly, the first direct proof of the very existence of atoms, predicted some 2,300 years earlier by the Greek scientist and philosopher Democritus.[1] This work eventually earned him his PhD. Another paper, based on what is known as the 'photoelectric effect', demonstrated that light, which was known to behave like a wave, could *also* behave as if it were a stream of particles, or small packets of energy, termed 'quanta'. This work was the very starting point of what was later developed into quantum theory, and was a fundamental first step towards the technological revolution of the 20th century. This paper eventually earned him a Nobel Prize.

Einstein 1904

The final two papers however, would change the world. *'On the Electrodynamics of Moving Bodies'* was a work based solely on thought experiments, the incredible product of a unique and brilliant imagination. It is an extraordinary scientific paper, containing no footnotes nor citations, and no mention of any work that had influenced or preceded it.[2] It became known as his 'special theory' of relativity. It was complemented, astoundingly, a few months later by a second paper, nothing more than a brief supplement, that contained without doubt the most famous, and amongst the most significant scientific equations of all time:

$$E = Mc^2$$

So what's so 'special' about special relativity? Unsurprisingly there is not one simple answer. A major reason was that it broke with a sacred set of laws that had underpinned science for the previous 200 years, those of no less a figure than the venerable Sir Isaac Newton. [3] Quite daring for an unknown 26 year old patent clerk, and most probably why it took another 15 years for his ideas to be generally accepted.

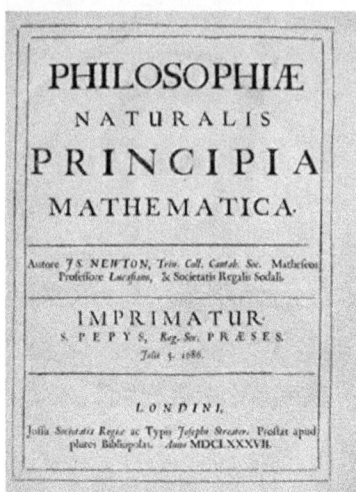

Sir Isaac Newton's 'Principia' 1687

Classical Newtonian physics is based on a framework of what is known as 'absolute' space and time, that is, in the Newton universe there is only one measuring stick, only one clock, and everything moves with reference to these. In the new Einstein universe however space and time were not absolute, but could be variable, taking different values for different observers depending on how they were moving relative to each other, hence 'relativity'. It was a universe where measuring sticks could change length, and where identical clocks could run at different speeds.

In the early 1900's all modern physics was based on the theories of the two fundamental forces then known to science: gravity as explained by Newton, and the electromagnetism of Maxwell. What Einstein had noticed was that the equations of Maxwell predicted, in all circumstances, a uniform and constant speed of light, which was completely at odds with Newtonian mechanics, where the speed of light could take different values depending on the motion of both source and observer. Clearly one theory or the other was erroneous. The popular story tells how the young Einstein, returning home one evening, was observing the famous Bern clock tower as he trundled along in a streetcar. As the clock receded into the distance, he pondered on what would happen if his streetcar suddenly sped up and started racing away at the speed of light. He imagined that to him, the clock would appear stopped, since the light reflected from the clock face would be unable to catch up with the streetcar. His own pocket watch however would be ticking away quite normally. Whatever the truth in the story, the point is that Einstein recognised that understanding the nature of light was fundamental in understanding the true relationship between space and time.

The wave-like nature of light had been well studied and firmly established since the early 1800's. However if light were a wave, then surely it should be a wave *in* something, rather like waves on the surface of a pond are waves *in* water. That something was popularly known as the *luminiferous aether*, a sort of elastic like medium presumed to fill the whole universe and through which the light waves could travel. It was an idea that had been around for centuries, and a concept universally accepted by the scientific community, from Newton to Descartes to all the leading figures of the day.

Enter stage an American scientist from Cleveland Ohio named Albert Michelson, who in the early 1880's set out with his chemist friend Edward Morley to make a series of painstaking measurements to demonstrate the existence of this aether. In a now famous set of observations using direct sunlight at different times of the year, Michelson expected to find small differences in the speed of light depending on how the Earth was at that moment flying through the aether with respect to the Sun. The idea was that waves of light in the aether should be rather like waves on the sea, which when viewed from the deck of a ship would be measured traveling at different speeds depending on if the vessel was heading directly into the waves, or running with them in the same direction. To make these very precise measurements Michelson had developed a novel piece of kit called an interferometer, the building of which had been financed by Alexander Graham Bell, the famous inventor of the telephone. The work was exhausting, and took more than half the decade to complete, but by 1887 they had their measurements.

The result was one of the most famous 'negatives' in the history of science. What they measured was that the speed of light was exactly the same, in all directions, and at all times of the year. In other words, there was absolutely no evidence for the aether, and the constant speed of light was at odds with Newtonian physics but completely consistent with the electromagnetic theories of Maxwell.[4] For his work, Michelson became the first American physicist to receive a Nobel Prize, and for Einstein it was the key ingredient in formulating his new theories. He did away with the need for an aether, embraced the concept of one universal speed of light, and as a result was able to show that neither time nor space were absolute. He could demonstrate, at least mathematically, phenomena such as that of a clock, when observed traveling at speed, would mark time *slower* than an identical stationary one. This effect we know as 'time dilation', where the greater the speed, the slower the clock appears to run. Also he showed that an object, again traveling at speed, would appear to a stationary observer somehow shortened, or *contracted*, an effect termed 'length contraction', with the greater the speed, the greater the contraction.

Stationary clock   Moving clock

In 1977 an experiment was carried out at the CERN laboratory in Geneva, Switzerland, where tiny *muon* particles were accelerated around a circular track until they reached a velocity equivalent to just over 99,9% the speed of light c. Under normal conditions, muons don't hang around much, decaying in about 2.2 millionths of a second. The speeding muons however were observed to have a lifetime more than 29 times that of their stationary brothers, the exact results agreeing perfectly with Einstein's equations. Similar experiments were carried out in 1971, using highly accurate atomic clocks flown at high speeds and compared to identical clocks upon return. Again, a 'time dilation' was observed in complete agreement with the special theory.[5]

To give Newton his due, it is not that his equations can be considered wrong, quite the contrary, they have served, and continue to serve us very well. Since we live in a world where things move extremely slowly compared to light, the effects of relativity are imperceptible. A person spending his whole working life driving high speed trains will only be a fraction of a second 'younger' than his office bound neighbour by the time he retires. But the effects of relativity are there, they are real, they are measurable, and they are a critical first step to understanding how space, time, mass and energy all interact to give us the universe we live in. The next step, and the sting in the tail of special relativity, was in the last paper, that little equation.

$E=Mc^2$

"I know not with what weapons World War III will be fought, but World War IV will be fought with sticks and stones": Albert Einstein

The Marshall Islands are a small group of islands in the South Pacific, about half way between Hawaii and Papua New Guinea. There, in 1952, American scientists exploded the world's first H bomb. 'Mike' as it was christened, exploded with a force of 10.4 megatons, the equivalent of more than 10 million tons of TNT, and some 700 times more powerful than the atomic bomb dropped on Hiroshima at the end of World War II. The mushroom cloud rose to 40km, and spread 160km across the sky. The huge blast was the result of a two part explosion. First a *fission* explosion, using tried and tested nuclear material, mostly uranium, was used as an initiator to get the party going. This was followed a split second later by the main *fusion* explosion using deuterium, a heavy version of hydrogen which exists naturally, for example in seawater. But where was all this explosive energy coming from?

First Hydrogen Bomb 1952

As Einstein put together his standard theory of relativity, showing how time dilated and space contracted, something else popped out: momentum, which you may recall from school days as being simply the product of mass x velocity. What his calculations on momentum were telling him was that as an object was given additional energy to make it go faster, it not only increased its velocity, it also started to increase its mass, first in very small amounts, but steadily increasing as the object got faster and faster. Eventually, as the velocity approximated that of light, practically all of the energy added would be going towards extra mass, and hardly any to increasing speed. At normal 'human' speeds, this mass increase is almost undetectable. However in a particle accelerator, such as the 3km straight tube at Stanford, California, electrons can reach speeds very close to that of light where their mass is seen to increase to around 40,000 times that of their normal rest mass.[1]

This understanding had two profound implications. First, it meant that nothing could travel faster than the speed of light, no matter how much energy in the form of rocket power or whatever else we gave it. There was, in effect, a 'speed limit' in the universe. The second implication was that the concepts of energy and mass were intimately bound together, as if they were interchangeable manifestations of the same thing, rather like water and ice are just different forms of H_2O. Of course the basic concept of mass and energy being interchangeable is in many ways extremely familiar to us, for example every time we eat. Just consider the weighty plate of pasta gobbled down by the marathon runner the day before a race.

So where we have mass we can have energy, and where we have energy we can have mass, and the two concepts Einstein related in his famous equation, with E being energy, m of course mass, and c, as we have seen, is the speed of light, which when multiplied by itself is a very large number indeed.

So if we have mass, we can have energy, but how much energy? Well, if we could convert *all* the mass of a substance instantly into energy, and we wished to create an explosion on the scale of say, H bomb Mike, all we would need is approximately 0.5kg of stuff, so something like a small bag of rice.[2] Of course complete and instant mass to energy conversion on this scale is not, as far as we know, possible. However we can create reactions whereby small amounts of mass are 'lost', and large amounts of energy released in the process. Nuclear fission is one, or energy released by 'splitting the atom' as it is popularly known. Fission relies on getting large unstable heavy atoms, like uranium, to split into smaller more stable elements, and in the process losing a tiny bit of their mass. This lost mass then appears as rather large amounts of energy in the form of heat and light, which under some specific conditions can give us a bomb, and under others, the chain reaction working mechanism of a modern nuclear power plant.

Another similar and potentially more interesting energy source is nuclear fusion, where instead of splitting heavy atoms, the process relies on 'fusing' together much smaller lighter atoms, once again losing a little bit of mass, and creating even larger amounts of energy. Nuclear fusion is the mechanism by which stars burn, where under the extremely hot conditions, hydrogen atoms are fused together to form helium, resulting in a loss of mass which appears as energy. Our own Sun for example burns every second 600 million tons of hydrogen, converting it into 596 million tons of helium, and 4 million tons of energy.[1] That's the energy equivalent of 8 billion H bomb Mike's exploding, *every second*. Up until 1905, physicists had been completely at a loss to explain the mechanism by which the sun could give off so much energy without burning itself up. $E=Mc^2$ solved that little problem quite nicely.

Nuclear fusion is also a possible future energy source here on Earth, with a fusion reactor having tremendous advantages compared to a traditional fission station. First, on a kilo for kilo basis, fusion releases huge amounts of energy, about 4 times as much energy as traditional fission, and nearly 4 million times more than carbon fuels. A typical coal fired power plant uses up about 2.7 million tons of coal per year, whilst a fusion reactor would need annually only around 250kg of fuel. Second, we have here on Earth an almost unlimited supply of that fuel. Deuterium, which is a 'heavy' form of hydrogen[3&4], would be the main source, and is present in sea water, about 33g per cubic meter on average, and we can calculate that just one litre of sea water could generate the same amount of energy as 300 litres of oil.[1] The fusion process would generate no carbon dioxide, with its main by product being helium, an inert gas used in children's party balloons, whilst radioactive materials left over would be minimal in quantities, relatively short lived and treatable. Finally with fusion there is no possibility of meltdowns nor chain reaction explosions, so no Chernobyl nor Fukushima-type accidents. Sounds too good to be true? Well, almost.

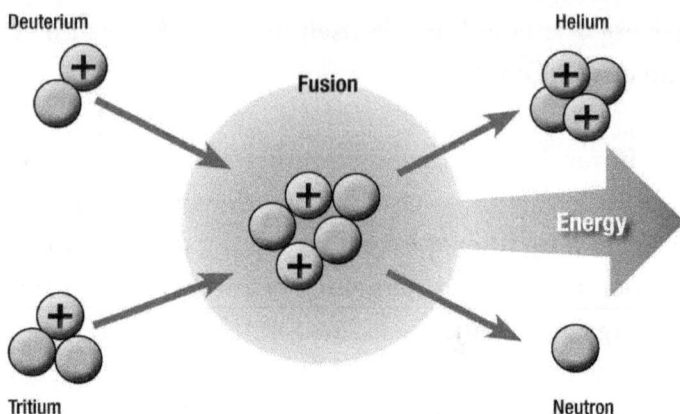

The technical challenge is enormous. A fusion reactor would need to operate at a temperature of 150 million degrees Celsius, or 10 times greater than the centre of the Sun, and would require enormous magnetic fields in order to hold the reacting plasma in place. However it is not impossible, controlled reactions have been carried out on a very small scale, and there is currently an exciting international project underway, ITER, with a much larger experimental reactor being built in the south of France, with the first plasma reactions projected for 2025.[5] Although the ITER project is not designed to produce energy industrially, it is hoped that it will give us the insights we need in order to subsequently design and manufacture a first generation of commercially viable fusion reactors, which with luck would turn nuclear fusion from its eternal ranking of tomorrow's technology into today's reality.

So $E=Mc^2$ kicked off an 'energy revolution', be it bombs, power plants, or simply enlightening us on the explosive natures of stars. However there was more. For example it also helped us to explain radioactivity, how a material could be constantly emitting energy in the form of radiation without melting away like ice. Using the theory we were able to understand that the energy came from unstable atoms inside the material, decaying spontaneously into lighter atoms, and releasing energy in the form of radiation, which again from school you may remember as alpha, beta and gamma rays. These insights in understanding the mechanisms of radiation were fundamental in its subsequent development in fields as diverse as medicine, geology and archaeology.

So in that one single year of 1905, the young Einstein had proved that atoms exist, set the quantum revolution in motion, explained how the stars are powered and completely changed the way we needed to look at the universe. However instead of being instantly promoted to Lionel Messi status, his work went largely unnoticed. He applied for a position as a university lecturer, but was rejected. He tried for a post as a high-school teacher, but was turned down.[6] Fortunately for the world, his physics continued. He knew that his standard theory was incomplete, something was missing, and that something was *gravity*. It took him a further ten years to complete the work, and in 1915, with the First World War raging across Europe, he published his masterpiece, his *General Theory of Relativity*.

Space Time

"Matter tells space how to curve, and space tells matter how to move": John Archibald Wheeler

During World War I, whilst Einstein was confined to Berlin due to his political opinions, a young British astronomer named Arthur Eddington held the post of Secretary of the Royal Astronomical Society. It was due to his position that he was one of the first scientists in England to receive letters and papers on Einstein's new *General Theory of Relativity*, which reached him through a mutual friend living in the neutral Netherlands. As luck would have it, Eddington was not only quick to grasp the significance of the work, but also sufficiently gifted as a mathematician to comprehend the finer details. Moreover, as a Quaker and a pacifist, he was open minded enough to undertake the challenge of promoting the research of a German scientist at such difficult times.

In early 1918, facing imprisonment for his stance as a conscientious objector to the war, Eddington was granted a reprieve on scientific grounds. Together with the Astronomer Royal Sir Frank Dyson, they had planned an astronomical expedition, whose aim was to gather scientific proof to support the new theory of Einstein. In early 1919 two teams were dispatched, one to Brazil, and the other, under the supervision of Eddington, to the island of Principe off the west coast of Africa. Their aim was to take astronomical readings during the total solar eclipse of 29th May, 1919, best observed from the latitude of their chosen sites.

Having shown in his standard theory that neither time nor distance were absolute as supposed by Newton, Einstein knew that the accepted theory of gravity would also need a 'relativistic' interpretation. Newton had recognized that gravity was a force intimately linked to the mass of objects, and his gravitational equations successfully predicted 'how' it worked, but his theory offered no explanation as to the 'why'. To Newton, gravity was some almost mystical force, some strange 'action at a distance' pulling objects towards each other across the voids of space. Einstein however knew that there must be a more fundamental explanation, and his 1915 paper was arguably the most brilliant game changer in the history of science.

Einstein of course was completely familiar with the work of Maxwell on electricity and magnetism, and the basic idea that there existed a field, the 'electromagnetic field', through which the forces of electricity and magnetism acted. He even saw the practical applications of this in the machinery and components that his family's company manufactured for the emerging power generation industry. What Einstein proposed was the existence of a similar field responsible for the gravitational force, a 'gravitational field', but with one big difference. Whilst the electromagnetic field was a field that diffused *through* space, the gravitational field according to Einstein *was the very fabric of space itself*. The vision he offered was of a space that was dynamic and variable, like a big sponge that could be squeezed and stretched, bent and twisted. This, in a nutshell, is the basic idea behind general relativity, and it had dramatic and profound consequences.

What Einstein saw was that any object of mass would curve and warp the space around it, with the bigger the mass, the greater the warping. A popular way of illustrating this idea of curved space is to imagine a saggy mattress, with a large heavy ball sitting in the middle. If we rolled a smaller ball across the mattress, its path would curve as it passed nearby the larger ball, or if it got too close and was going too slow, could even end up alongside it. Another visualization is to imagine that one morning, in your bathroom mirror, instead of the usual 'you' staring back, what you saw was yourself transformed into a distorted fairground caricature, with a big head, wide belly and stumpy legs. In the fairground of course this effect is created by using curved mirrors, however the same effect could be created if it were the actual space between you and the mirror that was curved.

In normal conditions the effects of space curving are very small, and very difficult to measure, even if extremely large objects are involved. When Einstein applied his ideas to a very large mass indeed, our Sun, he predicted that a beam of light from say a star, passing close by would behave in a similar way to the small ball on the mattress, and would be 'bent' as it followed the curved space around the Sun. To an observer on the Earth it would appear as if the star had moved a little from its normal position in the sky. The shift would be tiny, just a couple of arc seconds, but theoretically it could be measured. However since the brightness of the Sun makes it impossible to see starlight, the only way to observe and measure this effect would be during a total eclipse,

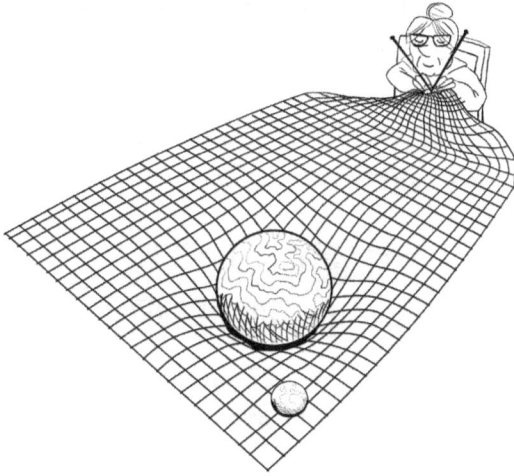

and it was this that Eddington and his team set out to achieve on the 1919 expedition.[1]

The Brazil team had bad luck with the weather, and neither was it looking good in Principe after a stormy morning, almost a case of rain stopped play! However it cleared a little in the afternoon, and a total of 16 photographs were taken in the few minutes the Sun was obscured by the Moon. The stars being observed were from a bright, well know constellation, and the quality of the images, although later a source of debate, were sufficient to make the calculations. It was confirmed. The measurements corresponded almost exactly to Einstein's prediction, a bending of the starlight by 1.75 arc seconds, a tiny amount, but enough to make history.

The results of the Eddington Dyson expedition made newspaper headlines around the world, and catapulted Einstein and his new theory into the public limelight where he would remain for the rest of his life. The New York Times report proudly announced that Einstein had somehow managed to get published a book that only 12 people in the world could comprehend, of course complete nonsense, but it did contribute to the aura of 'genius' that subsequently built up around Einstein.

LIGHTS ALL ASKEW IN THE HEAVENS

Men of Science More or Less Agog Over Results of Eclipse Observations.

EINSTEIN THEORY TRIUMPHS

Stars Not Where They Seemed or Were Calculated to be, but Nobody Need Worry.

A BOOK FOR 12 WISE MEN

No More in All the World Could Comprehend It, Said Einstein When His Daring Publishers Accepted It.

New York Times, November 10, 1919

Anyway, back to the physics. As far as the light from the star is concerned, it is not being bent by some strange force, but rather is simply following a natural path through space. It is the space itself which is bent, warped by the massive gravity of the Sun. Think Daytona 500, cars hurtling down the straight, the steeply banked curve at the end, and with little effort from the drivers, around they all go. Of course the physics is different, but the idea similar. Thus the Moon orbiting around the Earth, or the planets around the Sun, are in fact bodies moving quite freely through the curved space between them, curved space which they themselves have created. Matter is curving space, and space is guiding the matter, together in a sort of heavenly dance.

One of the most famous and dramatic examples of this was the flight of the ill-fated Apollo 13 mission in April, 1970, you know the one, "Houston, we've had a problem ..."[2] Just two days into the flight on the evening of April 13th (is 13 an unlucky number or what?) with the craft 300,000 km from Earth and heading towards the Moon, an explosion in one of the fuel tanks left the vehicle seriously damaged. It quickly became apparent that the goal of a Moon landing had to be abandoned, and all efforts focused on just trying to bring back safely the three man crew. As a result of the explosion, the crafts' energy reserves had been reduced dramatically, and if they had any chance at all of getting back to Earth, everything had to be shut down to an absolute minimum. Amongst the many problems they were facing was how to turn around the vehicle and get it back on course for Earth whilst preserving enough energy get it through re-entry. Mission control in Houston did the calculations and quickly rejected the most obvious option of a quick turnaround, as the fuel burn required to do a 180° in mid-flight and fire it back towards the Earth was just too great.

Plan B was to put the ship into what is known as a 'free-return' trajectory, effectively using the gravity of the Moon to sling-shot the injured craft back towards the Earth. So just five hours after the explosion, a brief 30 second engine firing modified slightly the trajectory, and Apollo 13 sailed on into the vicinity of the Moon where the curved space freewheeled them into an orbital path around the 'dark side'.[3] As they came out of this orbit, the engines were fired for a second time, just long enough to gain additional speed and shoot them on a course for home. The second burn was much longer than the first, almost 5 minutes, since they had to reach a speed sufficient to get them fully out of the Moons gravitational pull, and to the point where the Earths field took over.[4] You can almost picture the spacecraft as if it were a truck out of gasoline, having given itself a last blast of speed just sufficient to make it up and over the hill, before being able to freewheel down the far side, picking up speed, until it reached the nearest gas station. Apollo 13 made it to its gas station, and splashed down on 17th April in the middle of the Pacific Ocean with an exhausted, but alive, crew.

Another almost immediate proof of general theory was that it managed to solve at a stroke a long standing astronomical conundrum related to the strange orbital oscillations of the planet Mercury, which for many years had been seen by astronomers to not follow exactly the predictions given by the equations of Newton. With the new theory however the observations were perfectly explained, and it served to help convince a doubting world that Einstein was indeed on the right track.

As well as distorting space, general relativity predicted that mass would also distort time. The greater the mass, the greater the gravity …. and the greater the gravity, the greater the distortion of time. In his standard theory of 1905, Einstein had shown that a clock traveling at speed would appear to go more slowly, or lose time, compared to an identical stationary clock. When gravity was introduced into the equations, a similar effect was predicted. The greater and more intense the gravity field, the slower the clock would go when compared to a similar clock further away in a lesser gravitational field. Or to put it the other way around, from our Earth bound standpoint, *time runs faster upstairs*. In our normal day to day life these effects are of course way too small to be noticed, nonetheless they are there, and they can be measured.[1]

One important proof and application of these tiny time differences predicted by relativity is the worldwide GPS[5] navigational system, which each and every one of us uses almost daily. Initially developed in the 1970's for military and aviation use, the system relies on a network of 24 satellites orbiting about 20,000km above the Earth's surface, each containing an atomic clock constantly sending out signals which have coded into them the exact time they were sent. When the signals are picked up by, say, our mobile phone, it can use these time encoded signals to calculate our whereabouts very accurately. Indeed, to calculate our position to within 15m, everything needs to be in synch to an accuracy of 50 nanoseconds, or 50 billionths of a second.[6]

Now here's where relativity comes in. We know from special relativity that in the satellite the atomic clock will be for us running slower due to the high speed it is flying through space, about 7 microseconds (millionths of a second) slower per day. However the satellite is also 'upstairs', in a lower gravitational field, where according to Einstein the time is running faster, about 45 microseconds faster every day. The net effect is that the satellite clock will be gaining on our mobile phone by 38 microseconds per day. Well, it doesn't sound much, but if the system were not adjusted for these tiny differences, it would cause navigational errors that accumulate faster than 10km per day.[7] Or put another way, if it were not for Einstein, the whole GPS system would be failing after about 2 minutes. So the next time you are connected to Google Maps, making your way to that pub rendezvous, have your first beer in his honour.

Clocks running at different speeds often prompts the question about the possibility of time travel. Unfortunately as far as we know in our universe this is not possible, and if it were, it would clearly violate both the standard and the general theories of relativity. That said, it is theoretically possible to *travel forward in someone else's time frame*. Imagine that you had the possibility of flying from Earth to spend a few days holiday hovering around in a massive gravitational field, like those found around the edges of a black hole for example.[8] As long as you managed to not fall in, you could enjoy the sights for a while only to find on returning home that you had been missing weeks, months, or ever years, since the clocks back home would have been running much faster than yours. However even in this improbable scenario you would not have travelled forward in *your own time frame*, since your clock and your body and your brain would all have behaved exactly as normal. As for traveling the other way, backwards in time, best to leave that to the sci-fi enthusiasts.

Another exotic and popular idea, which this time is completely consistent with general relativity, is the idea of 'wormholes', or to give them their correct name, Einstein-Rossen bridges. The basic idea is that in parts of space so warped and bent by gravity, there could exist some type of 'short-cut' or hole giving access and quick passage to a completely different part of the universe. It can be imagined as two points marked on a flat piece of paper, where the shortest line between them is a straight line. Bend the sheet back on itself however, and the shortest line would be to pop directly through the gap in-between. However before you get excited, it is sufficient to say that these are only hypothetical, have never been detected, and even if they did exist and the mathematics were correct, then they would be so extremely tiny, of the order of 10^{-33}cm, that only very tiny worms indeed could hope to make it to the other side.

One of the strangest predictions resulting from general relativity was that of the existence of gravitational waves, although even Einstein doubted that they could ever be detected here on Earth due to the extremely tiny effects they would produce. Gravitational waves we can imagine as ripples on a pond, generated by some massive cosmic event, and spreading through space at the speed of light, squeezing and stretching everything in their path, gradually getting weaker and weaker. In 2016 the international LIGO[9] Scientific Collaboration group published results demonstrating for the first time the detection of such waves, and in 2017 the team were awarded the Nobel Prize for their work. Detectors had been set up at various sites in the USA and Europe, working on the same interferometer idea used by

Michelson and Morley back in the 1880's. The typical instrument used consisted of lasers and mirrors separated various km's along detector 'arms', which when exposed to passing gravitational waves could undergo very tiny, but detectable, changes of length. By tiny changes, we mean really tiny changes, of the order of less than a thousandth of the diameter of a single proton particle.

The first wave detected lasted for a mere blip of 0,2 seconds, and was named GW150914, i.e. Gravitational Wave of 14 September 2015, almost 100 years to the day since the publication of general relativity. It was a shock wave emitted from a binary black hole system more than 1.3 billion light years away, with the two black hole spiralling into each other, and the final event sending out a gravitational wave so massive it was just large enough to be detectable here on Earth 1.3 billion years later. As improbable as it all sounds, statistical analysis of the data collected over the period of detection were assigned a confidence level of 99,99994%, and further detections from other black hole systems were subsequently made over the next few of years. As well as providing a final confirmation, if one were needed, of the general theory of relativity, it also established an additional form of detecting and collecting cosmological information. Traditional telescopes collect and analyse electromagnetic waves of various frequencies, giving us light telescopes, radio telescopes, even x-ray and microwave telescopes, to which we have now added a new generation of gravitational wave detectors.

One final consequence of the general theory that fell out of the equations, was that space was not standing still. Indeed if the ideas were correct, it predicted that the universe must be dynamic, either expanding or contracting. Einstein initially did not like at all this idea, and even went so far as to introduce into his equations a 'cosmological constant', a sort of mathematical fudge to eliminate this prediction and leave behind a nice, stable, static universe. However over the years as the evidence built, Einstein came to recognise this as one of his most unfortunate errors, his *"biggest blunder"*, and the cosmological constant discretely disappeared.

At the end of the 19th century, science was fairly happy with itself. All the basic laws of nature were thought well understood, and the future appeared to be just a case of tying up the loose ends. But then along came young Albert Einstein, and while nobody was looking started to rewrite the rule book. The next lot however would tear it up into tiny little pieces, quantum pieces.

Waves are Particles

"Not only is the universe stranger than we think, it is stranger than we can think." Werner Heisenberg

When Winston Churchill made his famous remark in a 1939 radio interview about Russia being *"… a riddle, wrapped in a mystery, inside an enigma"* he might easily have been talking about quantum physics. For most people, the very word 'quantum' conjures up even today an image of some incomprehensible 'high science', something at the cutting edge, shrouded in a strangeness impenetrable to all but the gifted few. However, whilst it is certainly true that many of the ideas are extremely counter-intuitive, and that the underlying mathematics are indeed a jungle, the basic concepts at least are well within our grasp.

Over the next two chapters we will follow a little of the story that took us to our understanding of the structure of the atom, and the nature of its constituent parts. As the picture built, physicist observed that in this unimaginably small world, things behaved completely differently than they did in our everyday macro environment. Indeed so different was this new 'quantum' world, that it became apparent we had neither the verbal nor indeed the mathematical language to describe it. Waves behaved like particles, particles behaved like waves, and everything danced to a random tune. To get to grips with all of this, physicist had to put together a whole new and extremely complicated branch of mathematics based on probabilities and wave-like description of reality. And there, in essence, we have it: *quantum* is the tiny world of the atomic and sub-atomic; *quantum mechanics* are the mathematics developed to help us comprehend it. However, how we got there is a story worth telling, and like all good stories, we should start at the beginning.

The International Solvay Institute for Physics and Chemistry is located in Brussels, and it was here in 1911 that the first international conference was held, dedicated to advancing important areas of research in both fields. The fifth conference of 1927, on *Electrons and Protons*, was without doubt one of the most celebrated get-togethers in the history of science. Over six days some of the greatest ever figures met under one roof to discuss the newly formulated quantum theory. Of the 29 participants, 17 were, or were to become, winners of the Nobel Prize, amongst them household names such as Max Planck, Marie Curie, and of course, Albert Einstein. The historical group photograph is probably the most famous in all of science, and if there exists a sporting equivalent, it could be, at least if you are into basketball, perhaps that of the 1992 US Olympic 'Dream Team', of Bird, Pippin, Jordan, Barkley and 'Magic' Johnson. This was the golden age of physics.

1927 Solvay Congress Conference on Electrons and Photons

A. Piccard, E. Henriot, P. Ehrenfest, E. Herzen, Th. De Donder, E. Schrödinger, J.E. Verschaffelt, W. Pauli, W. Heisenberg, R.H. Fowler, L. Brillouin;
P. Debye, M. Knudsen, W.L. Bragg, H.A. Kramers, P.A.M. Dirac, A.H. Compton, L. de Broglie, M. Born, N. Bohr;
Langmuir, M. Planck, M. Curie, H.A. Lorentz, A. Einstein, P. Langevin, Ch. E. Guye, C.T.R. Wilson, O.W. Richardson

1927 Solvay Conference

By the time of the 1927 conference, the quantum revolution was well underway, and the participants were trying to come to terms with the strange and unusual ideas confronting them. Right up to the end of the 19th century, the world was governed by the 'classical' physics of Newton's gravity, Maxwell's electromagnetism, and all that was built on these foundations. It was a world of clarity, based on direct experiences, observable and measurable quantities, and predictable and intuitive processes. However cracks were showing.

It is worth our while at this point to take a small detour to mention the 19th century contributions of physicists such as Oersted, Faraday, Hertz and in particular the Edinburgh born mathematician James Clerk Maxwell, and the work they did on electricity and magnetism. After gravity, these two 'new forces' were correctly seen as fundamental forces in nature. Initially they were thought to be independent of each other, however so strongly were they interlinked that they were eventually shown to be manifestations of the one same force, *electromagnetism*. The complete theory and the mathematics that went with it were brought together by Maxwell in his 1873 paper *Treatise on Electricity and Magnetism,* one of the most significant works of the 19th century, still relevant today, and considered by many to be on a par with the famous *Principia* of Newton. It was the work of these scientists combined with the inventiveness of engineers and industrialists such as Edison, Tesla and Westinghouse at the very end of the century that brought electricity into our lives, and lit up our world.[1] Something else that fell out of the new theories of electromagnetism was an explanation of the nature of light. Since the early 1800's it had been known that light was wave-like. What was added was that light waves were actually *electromagnetic waves*, fluctuations or oscillations in this new *electromagnetic field.*

Around the same time that Michelson and Morley were doing away with the famous luminiferous aether, a German scientist called Max Planck, working in the rather obscure area of statistical thermodynamics, was grappling with another problem that classical ideas were unable to explain, which related to the way that bodies absorbed and emitted energy in the form of electromagnetic radiation. This radiation, similar to light, was classically expected to seep in and out of a body in a continuous wave-like way, rather like water soaking in and out of a sponge. However this idea did not correspond at all to Planck's measurements, and the solution he proposed, whilst a bit of a fudge, was quite ingenious. He supposed that the energy was being absorbed and emitted not as a continuous wave, but rather as *very tiny packets of energy*, which became known as *quanta*.[2] These quanta of energy really were extremely tiny, with the energy of each related to its frequency through a new constant h, which inevitably became known as Planck's constant, a number so extremely small it has 33 zeros in front of the first digit.[3] It was pure conjecture, but Planck's interpretation fitted with the experimental results perfectly, and in 1918 his contribution finally earned him a Nobel Prize.

In the years that followed, Planck's constant was seen as being absolutely fundamental in the unimaginably small world of quantum mechanics, popping up in all sorts of important theories and equations. We can even appreciate it in our everyday lives, for example in the colour of our Sun, where together with the temperature, it sets the colour of the light that reaches us. If h were, for example, just 25% smaller, then instead of its usual red yellow, our Sun would be an odd combination of purple and violet.

$$h = 0{,}00000000000000000000000000000000006624 \text{ Js}$$

Another interesting concept associated with the constant h is the Planck length. This is the hypothetical length below which it is impossible to determine the position of an object to any greater precision, and consequently the point where the very concept of length loses meaning, so setting a sort of minimum scale to the quantum world.[4]

The Planck length could also be a possible response to one of Zeno's paradoxes. These were a set of famous philosophical problems set around 450BC by the Greek philosopher Zeno of Elea to ridicule a school of thought supported by the rationalists of the day. One of the most well-known related how the swift footed Achilles challenged a tortoise to a race, naturally giving him a sporting head start. Achilles quickly reached the starting point of the tortoise, but of course in the time it took him to get there, the animal had moved a little further along. Achilles continued, but when he arrived at this second place, again the tortoise had moved forward some short distance, and so on and so forth, the point being that Achilles never quite catches up with the tortoise. This paradox has many interpretations, but one reflection is that you just cannot keep dividing things up into infinitely small quantities. There comes a point where it just makes no sense, and in our universe at least, it seems that for distances, the scale set by Planck might well be it. In essence, we begin to get a feel for what is at the very heart of our understanding when we talk about 'quanta', which is the idea that somehow nature at its very limits is 'granular', that the continuous gives way to the discreet: to the discreet 'quantum' packets of energy envisaged by Planck, or to the 'quantization' of space itself suggested by the Planck length.[5]

In 1905, in his famous paper on the photoelectric effect, Einstein found himself looking at something very similar to Planck, and so as it turned out, was the solution. The photoelectric effect is one whereby the energy of light falling onto certain materials can kick out electrons.[6] Just like Planck, what Einstein observed was that the way the light and the electrons exchanged energy made no sense if light was considered as a continuous wave. It did however make perfect sense if the light was thought of as small particles, or individual packets of energy, i.e. the same quanta envisaged by Planck. These 'blobs of light' were given the name photons[7], and what Einstein was proposing was that light, which for over 100 years had been clearly observed to be of a wave-like nature, was now also behaving as if it were a stream of particles. This conundrum became known as 'wave particle duality', a sort of uncomfortable co-existence of the two realities. It was as if light was made up of quanta packets so small that when they all lined up shoulder to shoulder, looked and behaved rather like a continuous wave.

So classical physics was on a bad run, but it was to get worse in 1911 when the focus switched to the work being done on atoms. Following the discovery of the electron by the British physicist JJ Thompson in 1897, the popular image that was emerging of the atom was one graphically explained as a 'plum pudding', a sort of hard solid mass with an overall positive charge, studded with smaller negatively charged electrons which had the ability to move and jump about. It was in his Manchester laboratory that the New Zealander Ernest Rutherford set up a now famous experiment, firing positively charged alpha particles at an extremely thin piece of gold foil. What Rutherford and his team observed was that whilst the majority of alpha particles sailed straight through as expected, a certain number were heavily deflected. It made no sense, almost as if a 15 inch naval shell had recoiled upon striking a piece of tissue paper Rutherford later commented.[8] When the mathematics were done, and the conclusions drawn, what they had discovered was the atomic nucleus, a small hard core to the atom where all the positive charge and practically all the mass was concentrated.

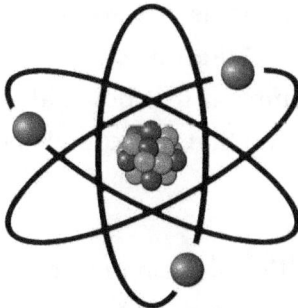

So the discovery of Rutherford gave a completely different picture of the atom, one of small negatively charged electrons 'orbiting' at a distance around a hard, much heavier positively charged nucleus of particles called protons, and it is this iconised picture which most of us carry in our heads even today.[9] One of the craziest aspects however, and completely wrong in our icon image, is the scale of things. The diameter of the whole atom is approximately 100,000 times greater than the diameter of the nucleus, or put another way, a relative volume ratio of 1.000 trillion to 1, so something like a pea on the centre spot of Wembley Stadium, with all the weight and all the energy being tightly packed into the pea. In other words, apart from the tiny nucleus, the atom is practically empty space, and lots of it.

However there were two problems associated with Rutherford's atom. The first was to understand how the positively charged protons, which naturally repelled each other, could all bundle up so closely into the extremely small nucleus.[10] An even bigger problem was related to the electrons, which if they really were orbiting around, should, according to Maxwell, radiate away their energy and collapse down into the nucleus, attracted by the positive protons. This would happen very quickly, in approximately 100 billionths of a second, and so in a flash our world would vanish. Quite obviously, and thankfully, this is not the case, which implied that down at the level of the atom, something was going on that was completely inexplicable using classical theories.

Two years later in 1913, a Danish physicist by the name of Niels Bohr, at that time also working in Manchester under Rutherford, came up with a proposal that at least gave us back a stable atom. The Bohr atom developed the quantum ideas of Planck and Einstein that down at these very small scales, the continuous made way for the discrete. He proposed that the electrons could exist around the nucleus only in certain fixed or discrete 'orbits', could not be found anywhere in-between, and could never fall below the lowest of these permitted orbits. These electron 'shells' as they became known, had the electrons moving only in specific energy levels, with the ability to 'leap' up or down from one energy level to another by absorbing or emitting discreet packets of energy, or quanta. This model at least offered a working solution, giving us into the bargain the much used popular expression a 'quantum leap'.

It also helped explain the mechanism of colour. When we shine light on something, the electrons absorb the energy in the light, and leap up one or more shells. When they fall back down, they re-emit energy in the form of the small blobs of light described by Einstein, the photons, with the exact amount of energy in each photon determining its colour, higher energy photons appearing blueish, and the lower energy ones reddish. Stuff is normally made up of combinations of many different types of atoms, with each atom emitting light of different colours, which when all combined, gives us the infinite array of colours that fill our world.

The same colour mechanism is also a useful tool to tell us what stars are made of. By analysing the light that reaches us, we can work out the atoms and the materials inside the star, the processes going on, how old the star is, and along with other data such as its size, even get an idea of how the star will evolve over time, either collapsing or exploding or simply burning itself out at the end of its life. Of even more significance however the Bohr model found great use in the field of chemistry, where it provided a completely new and solid theoretical framework for understanding the formation of molecules, and analysing chemical reactions, all in terms of interacting electron shells.

The contributions of Planck, Einstein and Bohr, and their use of this new concept of quanta were certainly successful in providing explanations to otherwise inexplicable phenomena, however these new ideas were completely at odds with classical ideas, and more insight was needed to make sense of this new quantum world. In 1916 Niels Bohr moved to the University of Copenhagen, where a few years later he founded the Institute for Theoretical Physics[11], part funded by the famous Danish beer company Carlsberg.[12] This now became the nerve centre, or world H.Q. of the quantum movement, where all the young guns of the day converged to confer and interchange ideas under the watchful eye of the father-like figure of Bohr.

The general theory of relativity offered by Einstein had been at best counter intuitive, and at worst downright complicated. What was about to be revealed by the emerging quantum theory however was just plain weird and spooky!

Particles are Waves

*"I do not like it, and I am sorry I ever had anything
to do with it": Erwin Schrödinger*

It was a young Frenchman by the very splendid name of Prince Louis de Broglie who got the ball rolling with his 1924 doctoral dissertation delivered at the University of Paris. His basic idea was that if waves can behave as particles, then cannot particles behave like waves? Sounds simple, although so revolutionary was his idea that the suspicious authorities had to consult with Einstein before awarding him his degree. Experimental research was soon underway, in the US by the team of Davisson and Germer, and in the UK by George Thomson, son of JJ Thomson, discoverer of the electron. Independently they demonstrated by the study of electron beams in a crystal lattice that indeed, the electron particles did behave with the wave-like properties envisaged by de Broglie. For his contributions de Broglie was awarded the Nobel Prize in 1929, and a few years later, in 1937, so were Davisson and Thomson. It is an interesting curiosity often commented that Thomson senior won his Nobel for showing that the electron is a particle, whilst Thomson junior for showing it to be a wave.[1]

Another famous experiment carried out at that time was the *double slit experiment*, where electrons were fired at a barrier with two small slits, and the emerging pattern recorded on a detecting screen placed behind. What emerged was a wave-like 'interference pattern', showing that the electrons were passing through the slits rather like an incoming wave from the open sea might pass simultaneously through two gaps in a harbour wall, recombining on the other side. Moreover, this happened even when the electrons were fired *one at a time*, so suggesting that even a single electron was somehow managing to pass through both slits *at the same time*. Whatever was going on, it was clear that the electron was not exactly the hard pebble-like particle physicists had thought it to be.

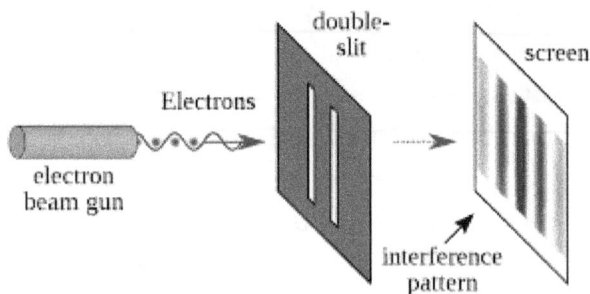

Double Slit Experiment

So if electrons were waves, then waves of what? And why did they behave sometimes like particles and sometimes like waves? At this point the world of physics went into a frenzy, very much in the hands of the young 'Sheldon's' of the day.[2] One was the German physicist Werner Heisenberg, who in 1925 aged only 24 came up with a new type of matrix mathematics to try to describe the unusual wave behaviour of electrons. Almost simultaneously, an Austrian colleague, Erwin Schrödinger, developed his famous 'wave equation' to explain things. What Schrödinger envisaged was that the electron was not a single point like entity, but was sort of 'smeared out' in a wave-like way. Then in 1927 Heisenberg added his equally famous *Uncertainty Principal* which postulated that we could never know with 100% certainty all the properties of an electron, such as its speed, position, momentum etc., which if true, added a certain 'fuzziness' to the whole quantum world.[3] Another German physicist, Max Born, brought probability to the party and suggested that the waves as described by Schrödinger were not really 'smeared out' electrons, but waves of probability, and that consequently we could never know exactly where an electron was, only the probability. It then took a brilliant Bristol born English mathematician, Paul Dirac, to add special relativity to the mix, and pull the whole thing together in his classic *Principles of Quantum Mechanics*, the theoretical mathematical basis for the whole new emerging field of quantum mechanics, first published in 1930 when Dirac was aged just 28. Take a deep breath. There will NOT be an exam at the end.

$$\left(i\hbar\gamma^{\mu}\nabla_{\mu} - mc\right)\psi = 0$$

Dirac's relativistic equation for the electron, combining the worlds of both 'h' and 'c'[4]

When the dust finally settled, needless to say Heisenberg, Schrödinger, Born and Dirac[4] all eventually received respectively a Nobel Prize for their contributions, and what remained was a picture of reality very different from the reality of our everyday experiences. Down at the atomic level things happened in a completely different way, much of which was becoming understood, although certain aspects were, and even today remain, without sufficient explanation. It is well beyond the scope of this small book to enter in any level of detail of the new quantum mechanics, but we can however try to give an overall taste of what emerged.

In quantum mechanics it appears that we can only know where a quantum particle is when it collides or interacts with something else, behaving in that exact instant like a particle. Between interactions however is a bit of a mystery, we have no way of knowing in a classical sense neither its 'flight path' nor where it will next pop up, and the best we can do is assign to it a series of probabilities in the form of a rather abstract mathematical equation known as a *wave function*.[5] One way to try to visualise this is to imagine you were in a dark London park, on a foggy Sherlock Holmes night, where you could only observe people as they passed directly below the dim gas lamps. You would see figures periodically appearing and disappearing at specific points, but in between you would have no real-world information on their whereabouts, and could only guess at what they were up to and where they were next likely to appear.

This conundrum is often summed up by saying that in the quantum world, things behave as particles when you are looking at them, and as waves when you are not. At the time this was a source of great debate, which was never really settled, until what eventually became known as *the Copenhagen interpretation* simply decided to accept this as a basic fact of quantum, and things moved on without much more ado. In other words, if the car works, then don't worry too much about what's under the bonnet. The *measurement problem* as it is known, to this day has no accepted explanation.

Another extremely odd feature of quantum is what is known as *superposition*. In our classical world, an object can either be in position A, or position B, but not both. In the quantum world however, a particle like an electron can be simultaneously in both A *and* B. This strange property of the quantum particle, to coexist in two places at once, only happens when it is in its 'unseen in-between' state, obeying the mathematical formulas of probability. This of course is the explanation of the double slit experiment, where a single electron is seen to pass simultaneously through both slits at the same time with a 50/50 probability. The strange phenomenon of superposition, as we will see later, is today being successfully exploited in the emerging field of quantum computing.

So if things down at the most fundamental level are governed by probabilities, why don't we see this in our everyday world? Why do I find my car in the garage every morning, exactly where I left it the previous evening, and not for example, occasionally outside on my neighbour's lawn? One way to think about this whole conundrum is to consider a casino. When a ball is rolled onto a roulette wheel, neither we nor indeed the casino manager have any idea where the ball will land, we only have a probability, 1 in 37. However over millions of rolls, the casino manager can build up a very accurate picture of where the ball, on average, will be landing, and of course can set the house odds in his favour. The point is that when we talk about the behaviour of one electron, we have a very fuzzy unclear picture, but if we consider complete macro systems of trillions of electrons that help in making you, me and everything around us, then we can have a very clear picture indeed.

This *probability interpretation* was hotly debated at the time, and was never really accepted neither by Einstein nor Schrödinger nor even Bohr. The very fact that at the smallest level there was no certainty, only probabilities, was the source of the famous comment by Einstein that he did not accept that *'God plays dice'*. It was with a similar view that Schrödinger devised his famous thought experiment of the cat in the box, whereby the poor animal was envisaged closed in together with a radioactive source, and a flask of toxic gas, which would be broken if it received some radiation. Since the radioactive decay necessary to break the flask is a quantum process, which according to theory is governed by probability, and therefore could be a simultaneous combination of both 'decayed' and 'not decayed', then we could never be sure without opening the box whether the cat was alive or dead. We would just have to accept, if superposition was a reality, that until we opened the lid the cat was also in a superposition of both dead *and* alive, something that seemed to Schrödinger and others quite ridiculous. Nowadays physicists would probably reply alluding to the incorrectness of mixing together quantum world and macro world states, however the objection does have a certain validity in the sense that exactly why the quantum world behaves in this strange probabilistic way is something else we don't fully comprehend.

From the work and equations of Dirac came another feature of quantum mechanics that originally sounded more like sci-fi, and that was the concept of *antimatter*. Dirac's equation for the electron worked, but appeared to have an inbuilt inconsistency in that it generated not only a solution for the negatively charged electron, but also one for an identical positively charged particle which was baptised the *positron*. Not only that, the equation also suggested that for every electron born, then an oppositely charged positron must also exist. At the time, this idea of an electron anti-particle seemed so weird that to many it simply suggested that Dirac's equation was incomplete. However Dirac stuck to his guns, and was eventually vindicated when, in 1932, Carl Anderson discovered the presence of positrons in cosmic rays as they travelled through the Earth's atmosphere. However there was more. Not only did Dirac's equation work for the electron, but it could also be applied to other fundamental particles, suggesting that basically every bit of matter had, somewhere in the universe, it's equal and opposite antimatter partner.[6]

Today antimatter is routinely observed here on Earth in the laboratory conditions of particle accelerators, and is an accepted part of our reality. It has also captured popular imagination due to its destructive capabilities, since if a particle of antimatter ever comes into contact with its normal matter equivalent, they are doomed to annihilate each other in an explosive burst of energy. If just 250g of matter came into contact with 250g of its antimatter relation, it would result in an explosion equivalent to 5,000 tonnes of TNT. So if one day you come across your anti-self, keep well clear![7] Another aspect of antimatter is the mystery surrounding its scarcity. If the theory is correct, both matter and antimatter should exist in equal quantities in the universe, although it appears that we actually live in a universe dominated by matter. Quite what has happened to all the antimatter is another of the unsolved mysteries of physics.

Another aspect that emerged, and without doubt the weirdest, is what is known in the quantum world as *entanglement*. As we have seen, quantum particles are governed by their wave equation, or wave function as it is known. Entanglement occurs when two particles somehow 'share' the same wave function. This can happen if the two are created simultaneously in some decay process for example. If we measure some property of one of the particles e.g. the way it is spinning, then we instantly know the same property of its twin. Also if we do something to affect one, then instantly we affect the other, as if the two are intimately connected even though they may be separated vast distances. Einstein called this *'spooky action at a distance'* and it was another idea of quantum theory with which he was extremely uncomfortable. However if he knew that this property of particles was being used today in experiments of *quantum teleportation,* well, it would have made his hair stand on end.

Teleportation in the popular sci-fi sense is about teleporting objects, i.e. the famous Star Trek "Beam me up Scottie". In the quantum sense however it is about teleporting information, passing characteristics from one quantum particle instantly to another. Much investigation has been carried out on quantum teleportation using photons of light, with one group led by the Austrian physicist Anton Zeilinger having managed to successfully teleport the properties of photons instantly across the 144km between the islands of La Palma and Tenerife in the Spanish Canary Islands.[8] More recent investigations appear to point to the possibility of quantum teleportation using electrons, which if viable, would help open up a wide array of possible future applications in the areas of quantum computing and quantum cryptography.

The American Richard Feynman, one of the most influential theoretical physicists of the second half of the 20th century, who in 1965 won his Nobel Prize for his contributions to developing the principles of quantum electrodynamics, is very often quoted as saying *"I think I can safely say that nobody understands quantum mechanics"*. Well, perhaps he was just being modest, because in reality we understand quantum theory very well indeed. The mathematical models developed by Dirac, Feynman and others are probably the most tested, incredibly accurate and predictive in the history of science. One example relates to a property of electron spin, known as the magnetic moment, where the value calculated using quantum theory equations has been experimentally proven to an accuracy of 12 significant figures, which is an accuracy equivalent to the width of a human hair over the distance between Barcelona and Paris. And it is through results like these that we have learned to apply quantum mechanics extremely successfully, in particular in relation to the behaviour of electrons and photons, and the subsequent development of the whole field of electronics and communication technologies which drive our modern world.

Despite its successes, there is nonetheless an underlying and uncomfortable feeling that certain very basic, fundamental concepts upon which quantum theory is based are far from fully understood, but in order to progress, we just accept them at face value and move on. We are like our friend the casino manager, who has no idea where the next roulette ball will end up, but who doesn't really care too much since he knows with complete certainty that over a busy weekend, he'll be in pocket!

Building Blocks

"Three quarks for Muster Mark!": James Joyce, Finnegan's Wake

So where did we get to with our atomic model? Ah yes, a fuzzy probabilistic cloud of negatively charged electrons, appearing and disappearing, surrounding a tiny but heavy nucleus of positively charged protons. But something was missing. James Chadwick, later Sir James, had studied under and worked with Rutherford both in Manchester and later at Cambridge. It was there at the Cavendish laboratory in 1932 that he led the team that discovered the *neutron*, a neutrally charged particle with a mass similar to that of the proton, with which it sat side by side in the nucleus. It had been more than 20 years since Rutherford had first discovered the nucleus, who at the time speculated on the existence of the neutron, but carrying no electric charge it proved dreadfully difficult to find. However it was probably just as well that it took a bit longer to come to light. Its small size, weighty mass and its inherent neutrality made it perfect for smashing into other nuclei, and of course, was the missing ingredient for provoking nuclear chain reactions. Had it been discovered earlier, it is highly likely that it would have put the possibility of developing a nuclear weapon within the grasp of the Third Reich, and the second half of the 20th century may have been very different indeed.[1] As it turned out, it was Chadwick who was later to write the final draft of the 1941 British MAUD report, which was fundamental in moving the US government to begin serious research on atomic weapons, and saw Chadwick himself recruited as head of the British team that formed part of the famous Manhattan Project.

So at this point things seemed quite clear. Everything was made of atoms, and these in turn were made of protons, neutrons and electrons. Add in the two fundamental forces of gravity and electromagnetism, and Hey Hoe! Together these few ingredients gave us all we needed to build the world around us, right? Wrong! As always, physicists were not satisfied with the picture, they wanted to probe even deeper, to see if there were yet more fundamental particles to be discovered, and to understand as completely as possible how everything fitted together.

In September 2008, a first proton beam was accelerated around the Large Hadron Collider at CERN on the French Swiss border of Geneva. The LHC is basically a circular tunnel 27km in circumference, 100m underground, through which runs a big old metal tube surrounded by superconducting magnets. The magnets in turn are surrounded by liquid helium to ensure they are kept at the necessary -271.3°C, just a little bit colder than outer space. Using the power of the magnets, beams of protons are accelerated around and around, whipped up to speeds close to the speed of light, completing 11,245 laps every second. A typical beam can consist of up to 3,000 bunches of protons, with more than 100 billion protons in each, and when at full speed will be traveling with the energy of a car moving at 1,600km/h. 2 Now here's the good bit. They then get a second beam whizzing around at the same speed in the opposite direction, and when the big boss presses the button ..... BAMM! Bits fly all over the place, and in a flash of a split second the wreckage photographed for an eager team of international scientists to pour over. And there you have it, a particle accelerator, or 'atom smasher'. Oh, and it cost €6 billion to build, and takes another €1 billion a year to run.

Large Hadron Collider, CERN

An intergovernmental meeting in Paris in December 1952 passed a resolution for establishing the *Conseil Européen pour la Recherche Nucléaire*, a European institution for nuclear research that was to *"have no concern with work for military requirements and the results of its experimental and theoretical work shall be published or otherwise made generally available"*. The first accelerator was up and running by 1957, and it has been one of the world's foremost nuclear research institutions ever since. Besides its important work in the field of particle physics, it is also credited with setting up in 1989 the world's first web page, and in the process basically inventing the World Wide Web, initially as a means of automated information sharing between scientists in universities and institutes around the world.[2] Throughout the 1950´s and 60's a lot of 'stuff' was discovered, and by the early 1970's physicist has more or less put all this together into what they called rather uncreatively the *Standard Model*.

New particles can be discovered in one of two ways. Either observing something new and trying to fit it to the theory, or more commonly, using the theory to predict that a certain particle can exist, and then looking for it. The basis of both methods takes us back to that little equation, $E=Mc^2$, which in turn explains why particle physics is such a complicated and expensive business. What Einstein's equation tells us is not only that where we have mass, we can have energy, so nuclear explosions and the rest, but also that where have energy, we can have mass, and if there is one thing that particle accelerators do have, it is energy, and lots of it. The result is that by creating these extremely energetic collisions, some of the energy involved can be transformed into mass, appearing in the form of new sub atomic particles that can be tracked, measured and given very strange names. The bigger the collider, the greater the energy, and the greater the energy, the bigger the possibility of generating new particles ...... and the bigger the € bill that goes with it. Whilst theoretical physicists need just a blackboard and chalk, experimental physicists need access to national, and often international, funding.

As well as strange names, most of the new sub atomic particles have something else in common, they generally don't like to hang around very much. The muon for example, a particle exactly like an electron just 200 times heavier, lasts about 2.2 millionth of a second. However two important particles were discovered that definitely do hang around, and are very much a part of our everyday world, and they belong to a group we call *quarks*. Proposed in 1964 independently by two US physicists, Murray Gell-Mann, and the Russian born George Zweig, they were eventually detected at the Stanford Linear Accelerator Centre in 1968. The two most basic quarks are the 'up' quark and the 'down' quark, and are the constituents of protons and neutrons. The proton is a combination of two up and one down quark, and the neutron two down and one up quark. Richard Feynman apparently suggested calling these particles 'partons', after the country singer Dolly Parton, but unfortunately it was Gell-Mann's proposal that stuck.[3] Why 'quark' nobody is sure, the only reference to the nonsensical word is in James Joyce´s last book, Finnegans Wake published in 1939. Perhaps it is because the book took 17 years to write, and is generally considered the most difficult work of fiction in the English language? Who knows, but along with the electrons, quarks were firmly established amongst the fundamental building blocks of our everyday world, and for his contributions Gell-Mann received the 1969 Nobel Prize.

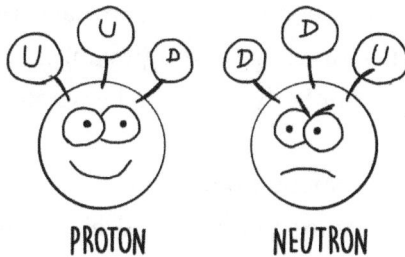

PROTON   NEUTRON

Another elementary particle present in extremely large numbers in our universe is the *neutrino*, meaning 'little neutral one'. It was first suggested in the early 1930's by the Austrian physicist Wolfgang Pauli, although it took until 1956 to be detected for the first time in a laboratory in the US.[4] The neutrino is not a very sociable particle, it carries no charge, has a mass a million times smaller than an electron, and can pass right through the whole Earth as if it simply did not exist. To detect it in its natural state, huge underground reservoirs of heavy water[5] had to be built, often in old abandoned mines, protected from unwanted background radiation. There the odd passing neutrino could interact with the heavy water, giving off characteristic tiny bursts of energy. The first neutrino detection using this method was in 1965 in South Africa, 3km underground in a specially prepared chamber in a gold mine near Boksburg.

Neutrinos are produced in huge quantities as a by-product of the nuclear processes in stars. They are so numerous, that after photons, are considered to be the second most numerous particle in the universe, with the combined mass of all the neutrinos thought to be similar to the total mass of all the visible stars. It is estimated that from our own Sun we receive about 65 billion of the things, *every second, every square centimetre!* However worry not dear reader, since even though you have right now many billions of neutrinos traveling thorough you, only one or perhaps two will react with your body atoms in the whole of your lifetime.

Something else fundamental discovered around this time were two new forces. Of course gravity and electromagnetism were well known, and to these were added the two forces that hold the particles and the atoms together: the strong nuclear and weak nuclear. The strong force as its name suggests is by far the strongest of all the forces, indeed it is the strongest known force in the universe, more than three trillion times stronger than gravity, and is responsible for holding together all the particles in the nucleus. However it has an extremely small range, explaining why the nuclei are so small, hard and compact. The weak nuclear force also acts over a very short range, and despite its name is still a couple of trillion times stronger than gravity, and is mainly involved in atomic decay processes like radioactivity, and importantly, in the fusion process of hydrogen into helium which powers stars.

So by the 1970's with their Standard Model physicists had pretty much summed up all of the particles and forces necessary to build nearly all of the universe as we know it:

Four particles: Electron; Up Quark; Down Quark; Neutrino

Four forces: Strong Nuclear; Weak Nuclear;
Electromagnetic; Gravity

You may, or may not, have been aware of a big commotion in the physics world a few years back, something about the discovery of a new particle, the *Higgs boson*? Well this really was a big deal. Despite all the successes of quantum theory and the Standard Model, there was one important piece of the jigsaw missing, which was an understanding of where the mass of particles comes from, and why some particles like the electron and the quarks have mass, whilst others such as the photon are massless. In the 1960's a theory was put forward that proposed a field to exist right throughout the universe, and the way in which particles did, or didn't, react with this field endowed them with their characteristics of mass. Conceptually, this *Higgs field* as it became known[6] can be envisage as a sort of background 'stickiness', interacting with some particles , whilst leaving others unaffected and free to zoom about unimpeded. Neat idea, but super difficult to detect. What physicists were looking for as a way of proof was the particle associated with the field, the so-called Higgs boson, and to do that, what was needed was huge amounts of energy, and lots of patience. Eventually this last missing piece of the Standard Model was discovered, in 2012, more than 40 years after it was predicted, and guess where? Yes you got it, the LHC at CERN.[7] Why was it such a big deal? Well, apart from being an important confirmation of the Standard Model, without the Higgs field, nothing would have mass, so electrons and all the other particles would be whizzing around at light speed, and there would be no atoms, no molecules, no chemistry nor biology, no you, no me, no world and no universe, at least as we know it. So yes, it was a big deal.

Standard Model of Elementary Particles

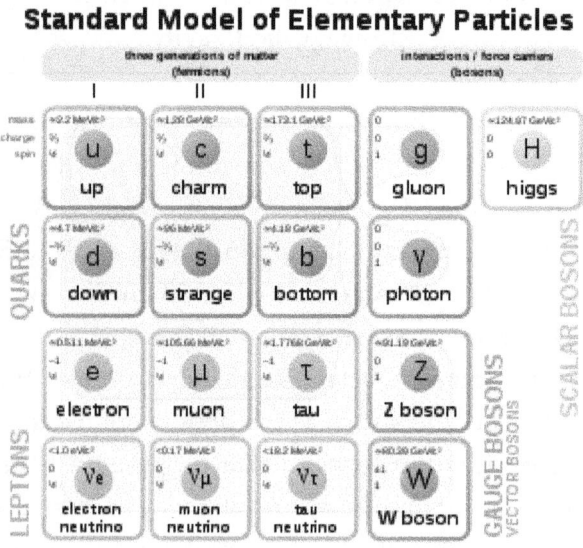

New Periodic Table

The schematic of the Standard Model is sometimes referred to by physicists as the 'new periodic table', which in retrospect is probably just to annoy chemists. However just like the periodic table of elements[8] it does show in a reasonably clear and organised way all the particles that we currently believe to exist. It turns out that the four particles already mentioned have some close relatives. The electron for example has a heavier cousin known as the muon, similar in almost every way except that the muon is heavier, appearing at much higher energy levels, and no sooner does it pop into existence than it disappears again. So fleeting is its existence, it never leaves the confines of the accelerator that produces it, and only crops up here on Earth naturally as a result of cosmic rays interacting with our atmosphere. Similarly the electron and muon have an even heavier distant cousin, the tau, and so on and so forth for the quarks and the neutrinos. We sometimes refer to these short lived 'heavier sets' as Generation II and III of the original set of four. Why do they occur? What role do they play? Basically we have no idea, we just know that they are there because we have seen them.

So the final picture we have today consists of 12 particles of mass, combined with the 4 massless particles of the forces, plus the Higgs boson, which, if you are counting, makes a grand total of 17 particles that really do appear to be the final building blocks of everything known to us. There is a hypothetical 18th particle associated with gravity, the 'graviton', but for now we have no experimental evidence as to its existence.

The quantum revolution has given us many wonderful revelations, but if we had to single out one, perhaps it is this, that after more than 2,000 years of science, we finally appear to have worked out what everything is made of!

Fields that Form the World

"Quantum Field Theory, which was born just fifty years ago
from the marriage of quantum mechanics with relativity, is
a beautiful but not very robust child": Steven Weinberg

So by the 1970's, the quantum story had brought us to a point where in the Standard Model we had a fantastic picture of the basic building blocks of the world around us, of all the different particles, and all the different forces that hold them together. However even as this model was being put together, it became apparent that there was something even more fundamental underpinning it all: *fields*.

The idea of fields in modern science started in the mid 1800's with Michael Faraday and his work on magnetism. Who amongst us hasn't been fascinated messing around with magnets, pushing together and breaking apart the opposing and attracting ends? There is nothing visible, but you can literally feel the forces in your fingers. Throw some iron filings about and you can even start to see the magnetic field, with the field lines showing the intensity and the direction of the field in the surrounding space. Similar fields are created by electricity, electric fields, and it was not long before Maxwell had bundled them together into one single theory describing the *electromagnetic field*.

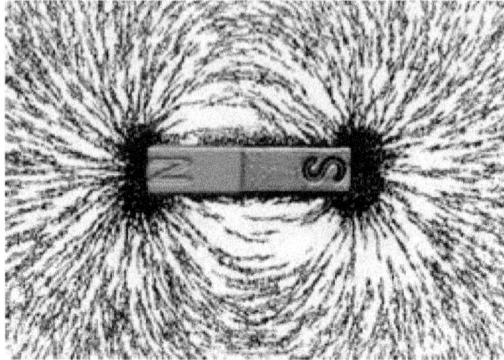

The definition of a field in science is something that has a certain measurable value, at every single point in the space through which the field is spread. Temperature in a room is an example, where we can measure the temperature at different points, and even plot the results to help us visualise the distribution, for example red for warmer areas and blue for colder. Gravity could be another, where we could measure not only the strength but also the direction of the gravitational force at any particular spot.

To visualise temperature distribution in a room is easy, but try to visualise the electromagnetic field and it gets very messy indeed. To start with we have light, which you will remember are oscillations in the electromagnetic field, traveling as small quanta wave packets, or blobs of light called photons. So unless we are sitting in a completely dark room, light will be bouncing around off every object, to every corner of the room, non-stop, photons everywhere, from the furniture, from the dog, from the kids messing about, even from the garden visible outside the window. Of course we can only directly perceive a very small fraction of all this by the photons that actually enter our eyes through our two tiny pupils, but in fact they are everywhere, and to check it out we just need to walk around the room. But light isn't the only thing making the electromagnetic field vibrate around us. We might have our mobile phone turned on, receiving text messages, picking up the internet or connected via Blue Tooth to a loudspeaker, all just more waves in the very same electromagnetic field, just at frequencies different from light. We can then turn on the radio or the TV, flip the channels and check out just how many signals from each and every station are also floating about, oscillating the same field, just waiting to be picked up by our devices. And we could go on, with cosmic rays, radar waves, and more. The point is that the one same electromagnetic field is all around us, everywhere, all the time, oscillating and vibrating and carrying unimaginable amounts of energy and information within it.[1]

Paul Dirac, 1933

Just as Newton had given us a classical theory of gravity, which eventually needed some rewriting from Einstein, so the classical electromagnetic theories of Maxwell had similarly to be updated and brought into the new quantum world. It was again the brilliant Paul Dirac who showed the way. Dirac took his mathematical models used to describe the quantum behaviour of electrons, and he applied them to the electromagnetic field, an idea that was developed further through the 50's and 60's by Feynman and others.[2] As physicists came to understand exactly how the electromagnetic field worked, how it interacted with the electrons and other particles, and how the same ideas could also be applied to the strong and weak nuclear forces, they came up with an idea so weird, so revolutionary, but eventually so successful that it has been the theoretical basis of the world of physics for more than 50 years: *Quantum Field Theory* (QFT).

The basic idea is that everything, absolutely everything, is made from fields, or more precisely, quantum fields. All the fundamental particles of our Standard Model are really just excitations in these underlying fields. Give a quantum field enough energy, a good 'shaking' as it were, and its particles will pop into existence. What exists on the very lowest, most basic level are first and foremost fields. There is one fundamental field associated with every fundamental particle, so if we have 17 particles, then underlying these are 17 fields. Each one of these fields extends right throughout our universe, and it's these fields which combine to make up you, me, the air we breathe, the food we eat, the car we drive, the sunlight touching our face, everything. Sorry, you were warned!

So let's take the electron as an example. According to QFT there exists, right throughout the universe, one field which we can call the *electron field*. Just like the simple temperature field we discussed earlier, this electron field will have a certain value, or energy, at every single point right throughout the whole of space. At most points the value, or energy, will be zero, or more specifically it will an extremely small set of tiny quantum fluctuation (consistent with the Heisenberg principle) around the point of zero energy. So if we created a complete vacuum, like we might find for example in outer space, with absolutely nothing it, no particles, no light, nothing, then there would still exist in that vacuum the electron field, shimmying around its zero point.[3]

Now let's shake things up a little bit and add some energy. Like plucking a guitar string, the added energy will cause vibrations in the field, and the resulting sets of tiny vibrations manifest themselves as the electrons so familiar to us. Consequently according to QFT the electron is nothing more than a local excitation in the fundamental underlying electron field, and all the electrons in the universe are part of the one same connected field. Of course the idea that electrons are the consequence of energy being given to the electron field is completely consistent with $E=Mc^2$, where we put in some energy E and create a mass m, in this case the rest mass of the electron.[4] Indeed this is the process we depend on in particle accelerators when we create new particles to study. Further, QFT implies that the very nature of the electron is not a hard pebble-like object, but rather a small set or packet of wave vibrations in the field, which of course is also consistent with the ideas of quantum theory.

The electromagnetic field behaves likewise, with the energy we give it 'appearing' as wave packet vibrations, or photons. Give it a lot of energy and we get a lot of vibrations in each wave packet, resulting in higher frequency photons like X-rays or ultra-violet light. Give it lower levels of energy and we get lower frequency photons, like infra-red light, microwaves, radio waves etc. And basically so on and so forth for all the other particles and forces in the Standard Model.

Another important characteristic is that these fields are all mixed up, sitting side by side and rubbing shoulders right throughout all of space, and so through proximity, energy vibrations in one field can kick off similar vibrations in another, which gives a working model for explaining how particles and forces interact, decay from one into another, and indeed all the processes we observe in our particle accelerators and in nature.

You may have noticed that our Theory of Everything equation of chapter one encapsulated all the basic concepts of physics, from forces, to particles, even to the Higgs boson, all as expressions of fields. Consequently what we are really considering as our Core Theory, is the world according to QFT, the Quantum Field Theory.[3] This theory has been unbelievably successful in forming our almost complete understanding of the world around us. However, as alluded to, it has its short falls.

The principle area of concern relates to the inclusion of gravity, which it would be amiss of us to gloss over. The basic conflict is that the mathematical descriptions of all the particles, and of all the other forces, are derived from quantum theory, and are wavelike descriptions of reality. The best theory we have of gravity on the other hand is Einstein's general theory, and the mathematics underpinning it are very much classical, and not in the least bit quantum. In most practical cases this is of little consequence, since they rarely overlap, with gravity explaining the very large where quantum effects are insignificant, and quantum the very small where gravity can be ignored. It is only when we get into areas where we simultaneously have to consider both that things get complicated and the maths breaks down, i.e. extreme situations, of tiny dimensions and massive gravity, like the singularity of black holes, or the first few instants of the Big Bang. It seems clear that both theories cannot be simultaneously true, and consequently the feeling is that there must be some sort of deeper more fundamental theory uniting them. A tremendous effort has gone on over the decades, and whole scientific careers have been dedicated to try to reconcile gravity with the other quantum forces, and to find some sort of underlying Grand Unified Theory (GUT). Even Einstein spent the latter years of his life working unsuccessfully in this area.

There has been two main approaches to GUT. The first has been to try to quantise general relativity, finding a model and the mathematics to make it consistent with the rest of quantum theory. One of the most popular and promising ideas in this area has been that of *Loop Quantum Gravity* (LQG).[5] Here the basic idea is similar to what we looked at when considering Zeno's tortoise paradox, the idea that space is not infinitely divisible, that ultimately it is 'granular' in nature, and that right down there at the smallest level it has a certain minimum value, something of the order of the Planck length. In LQG these minimum regions are envisaged as small 'rings' or 'loops', interlinked and woven together to give a dynamic chain mail mesh-like structure to the whole of space. For the time being however the ideas of LQG, attractive as they are, remain speculative, without any definable characteristic enabling us to detect or prove the correctness of the theory. One indicator would be finding the elusive graviton[6], although for the time being at least this appears beyond our capabilities.

The second approach to finding a GUT has been to define a completely new and different underlying theory, an alternative to QFT, one from which both gravity and quantum ideas can emerge naturally. Amongst the most popular of alternatives are the various versions of *String Theory*, where instead of a mess of multiple fields, what is proposed is just one basic underlying structure consisting of 'strings' of vibrating energy, filling the whole universe, with the way in which the strings vibrate giving us our complex world of particles and forces. It's a really cool idea, but again, at least for now without experimental evidence, and bringing with it its own messiness, including ten dimensional space, multiple universes, and such mind boggling ideas that, well, if you decide to go there make sure you have a good breakfast first.[7&8]

So for now, like we concluded in chapter one, QFT continues to be the very best we have, and safe in that knowledge you may drift off to sleep tonight, quietly reflecting that you are made up from a handful of fields, connected to all about you, and to everything else in the whole universe. May the force be with you!

Universal Measurements

"The most lasting and universal consequence of the French revolution is the metric system": Eric Hobsbawm

In June 1792, with the French revolution in full swing, two astronomers set out from Paris tasked with, quite literally, measuring the world. Under the charge and instructions of the *Académie royale des Sciences*, they were each equipped with a specially prepared carriage, the latest equipment and gadgets, as well as a couple of willing assistants. Jean-Baptiste-Joseph Delambre, then 43, was to make his way North to the coastal town of Dunkirk, whilst his colleague Pierre-Françoise-André Méchain, five year his senior, was to travel South to the Catalan city of Barcelona. Their mission, planned to take two years, was dogged by civil unrest and war almost from the outset, took seven years to complete, and their story amongst the most remarkable in the history of science.

Delambre and Méchain, both from humble backgrounds, formed part of the scientific establishment of the time, a group commonly known as the 'savants'. These pre-eminent thinkers were frustrated by the difficulties that they, and society in general encountered, living in a regime where there was no accepted unified system of weights and measures. Differences existed of course between nations, but even within France it was estimated that there existed close to 250,000 varying units of weights and measures. Even from trade to trade there were differences, with the weights of the baker being different to those of the silversmith, the lengths used by the builder different from those of the cloth merchant. Such differences hindered commerce, obstructed efficient administration of the state, as well as making life difficult for the savants to communicate and compare their investigations with colleagues and peers. They were the pioneers of the idea of a universal measurement system, a rational and coherent system for a rational and coherent world, which today we know commonly as the metric system. The revolutionary political winds of the time were also in their favour, with the idea of universal measurements striking a harmonious tone with the vision of universal rights for all, throwing off the shackles of custom, destroying local distinctions, and building a new and fairer world. In the same way as the Revolutionaries set about establishing a new linguistic standard and unity of the French language, eliminating regional languages and local dialects, so the savants saw their opportunity to establish their universal measurement system across all areas of scientific and public life. In the words Condorcet, one of the leaders of the movement, the metric system was to be *'for all people, for all time'*.[1&2]

Condorcet, Permanent Secretary of the Académie des Sciences, 1777-1793

Some of their ideas were in retrospect rather unusual. For example they proposed a new calendar, with year 1 starting on 22nd September 1792, the date of the founding of the first French republic, divided into 12 months, each of 30 days, made up of three weeks, or 'decades' of 10 days, with no Sundays nor saints' days. They proposed splitting each day into ten hours, and each hour into one hundred minutes, and even commissioned prototype clocks. They proposed splitting the circle into 4 quadrants each of 100° instead of 90°, with all the recalculations of the numerous mathematical tables that this implied. All of this was unsurprisingly to little avail, however it was in the area of weights and measures where their contributions were to have universal consequences.

The basic idea had three parts. The first was to have one interlinked system, where length, area, volume and weight were all of the same common standard. The second was decimalization, whereby the unit system would scale up and down in units of ten, something which seems obvious now, but at the time was not so clear, as alternative systems like those being used in the

US even today can testify. The final consideration was that the underlying definition of whatever units chosen should be based on a 'natural' definition, some sort of 'universal constant' that could be measured and contrasted independently, and not related to some subjective quantity such as the size of the Kings' 'foot', or the weight of a 'stone'. The unit of weight for example was to be based on something as common and universal as water, with the unit 'gram' to be defined as the weight of exactly one cubic centimetre of water. Similarly the unit of volume, the 'litre' would be the equivalent of one thousand cubic centimetres. Even the new currency, the franc, would form part of the system whereby one franc would be equal to 0.01 grams of gold, with one franc divisible into 100 centimes. As you may have noticed, the underlying unit of all these measurements is length, or the meter as it was to be called, and it was this that Monsieurs Delambre and Méchain set out to establish.

The savants had decided to use as the natural definition of their new meter nothing more nor less than the dimensions of the Earth, to be precise, one quarter of the circumference of the globe, or the distance from the North Pole to the equator, divided by 10 million. It was a bit of a fudge, since from previous studies the savants had a good idea of what the distance from pole to equator should be, and divided by 10 million would give a distance rather similar to the length measurement then being used in Paris, the *toise*. But like good scientists they needed it checked and double checked. Delambre and Méchain were to measure the longitude meridian of Paris, a straight line running from Dunkirk in the North to Barcelona in the South, passing through the middle of Paris. At both sea towns they could use astronomical measurements to work out the latitude of each extreme point of

their arc, then simply knowing the exact distance between the
two, they would be able to extrapolate their measured segment
of the orange to the full quarter Earth, which when divided by
10 million would give them their precise meter measurement. To
measure the distance between the two points, they used the tried
and tested method of multiple triangulations, with Delambre
working his way South, and Méchin North, with the idea of
meeting up somewhere in the middle.

Seven years and many adventures and mishaps later, the results
were presented to an International Commission in 1799, the
world's first international scientific conference, and the exact
measurement replicated in the form of a metal bar of pure platinum,
which from then onwards was kept under triple lock and key at

the National Archive in Paris. This was to be the world standard of the new meter, with copies being sent to the governments of other countries, and wooden meter sticks made by the tens of thousands to be shared amongst citizens throughout France. As a result of this work, the 'litre' and 'kilogram' were also defined with a similar platinum kilogram weight being manufactured and archived together with the meter.[3] Years later, with the benefit of space and satellite technology, the exact distance of the meridian from pole to equator has been measured at 10,002,290 meters, meaning the meter calculation made by Delambre and Méchin was short by 0,2mm, or about the thickness of two sheets of paper, simply remarkable given their task and the circumstances under which they had to work.

With such a history, it is of no surprise that today the world HQ for scientific measurement standards is based in Paris, with the system used known as the SI unit system, from the French *Système International*, a body established in 1960. Of course it is no longer a purely French driven affair, being in reality an international scientific forum, whose ongoing work is the definition of international measurements, standards and accuracies. The basic idea of the whole SI system remains very much that of the founding fathers, the savants, i.e. that the system should be interlinked, metric, and based on naturally occurring and measurable constants. The SI system is built on seven 'base units', and it is from these seven that all other measurement units of science can be derived, so for example for length we have the meter, for time the second[4], and for weight the gram, or more popularly, the kilogram. In turn, each of these is defined in terms of some naturally occurring constant. In the case of the meter, it is now defined in terms of the velocity of light c, so there is no need for scientists to be running back and forth to the National Archive in Paris to check their measurements against the platinum stick. In 2019 there was some excitement in the scientific community when the final base unit which until that point was still defined in terms of the Paris platinum standard, the kilogram, was finally given a definition in terms of another of our favourite constants of nature, the Planck constant h.

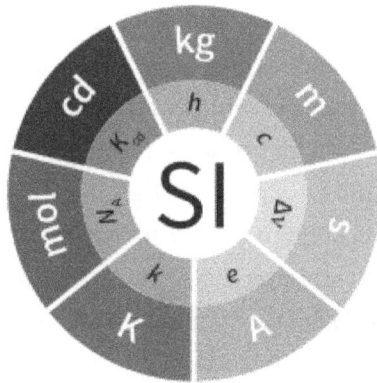

So Condorcet and his colleagues of the Revolution would no doubt be well happy with the modern state of affairs regarding scientific standards and measurements. At least in the scientific and international world of trade and commerce, the system is now universally accepted, and in the vast majority of countries worldwide it is the chosen system of daily use. The British imperial and U.S. customary measurement systems have been historically the most widely used alternatives, although even they are now legally defined in terms of SI system equivalents. For their part the British part-decimalized in the early 1970's, coinciding with their entry into the European Common Market, although even today certain imperial measurements such as gallons, miles, pints, feet and stones remain in popular use. In the US decimalization has been debated since the times of Washington and Jefferson, although being always politically closer to the UK than to France, it is of no surprise that they chose to adopt the British system, and remain today the major international exception.

Working outside of the SI system can have its drawbacks however, as the U.S. discovered to its cost in 1999 with the loss of the Mars Climate Orbiter. The craft, after a 10 month journey to reach the red planet, crashed and was lost due to a trajectory error. The cause was eventually traced to the engineers at Lockheed having used imperial units in the vehicle design and performance specifications, whilst the flight planners at JPL had used metric for all of their trajectory calculations. Between the two, something got 'lost in translation', resulting in a $125 million embarrassment for NASA, and a lesson on the value of universal measurements. Thus even in the U.S. the SI system is generally now the scientific norm, and it is through the field of science and technology that the 'silent revolution' of the savants continues.

Quantum Computing

*"Computers of the future may weigh no more than
1.5 tons": Popular Mechanics magazine, 1949*

The quote heading this chapter could easily have been substituted by other similar futuristic pearls of wisdom, such as that of Thomas Watson, Chairman of IBM who in 1943 was credited with the comment that he believed there was "a world market for maybe five computers". Or the vision of Ken Olsen, President of Digital Equipment Corp who even as late as 1977 could see "no reason anyone in the right state of mind would want a computer in their home". Hindsight is a wonderful gift. The point is that whatever we write about computers today will almost certainly be out of date by the time it is read. The best we can hope for is to stick to a few essentials.

If there is one single notion about quantum computing worth taking away which won't change over time, it is this: that quantum computers are not simply smarter, faster, more powerful versions of our current friendly digital microprocessors, but represent a completely new and different technology which we are only just beginning to explore. Quantum computing is to digital computing what the electric light bulb was to the candle.

The notion we have of our computers is that they are smart. This is false, they are dumb. Dumb, but fast.[1] All traditional computers, from the very early valve versions which filled whole rooms, to our friendly lap tops are so dumb that they can only count to 1, or more correctly they can distinguish between a 1 and a 0, but not much else. The trick is that they can do this with extraordinary speed and perfect accuracy. Over the years, as our technologies have advanced we have given them 'memories', electronic filing systems where they can store and retrieve strings of 1's and 0's, and this capability has in turn enabled us to get very clever with our programs and algorithms, allowing us to assign to our computers ever more complex tasks. But at the end of the day, what they do, is crunch 1's and 0's. Even in 1996 when IBM's Big Blue computer trounced the world's chess champion Gary Kasparov, it was not through a superior intelligence, but rather by the application of sheer mindless brute force, crunching through millions of combinations in fractions of a second, and employing its mega memory and clever programming to call the plays.

Richard Feynman 1959

So what's the big idea about quantum computing? And where does all the talk of 'computing power' come from? It is often cited that one of the earliest proponents of quantum computing was none other than Nobel prize physicist Richard Feynman, who at a conference organised between IBM and MIT at Endicott house, Boston in May 1981 famously postulated that *"if you want to make a simulation of nature, you'd better make it quantum mechanical"*. What Feynman was getting at was that nature, at its very heart, is quantum, and as a consequence, of an exponential complexity such that to model it we would need exponential computing solutions.

The idea of the exponential is fundamental to understanding the field of quantum computing. A simple example is to consider a dinner party of 10 people, where the number of different possible seating plans is factorial 10, so 10 x 9 x 8 x 7 .... = 3,628,800. Every time we add just one new person, the number of possibilities just shoots up. Imagine arranging the seating plan for a whole wedding! Another famous example tells the tale of the creator of the game of chess, who took his idea to the Emperor of ancient China. Being suitably impressed, the Emperor asked what he wanted in return. Humbly, the inventor looked at the chess board with its 64 squares, and asked the Emperor only that corresponding to the first square on the board, he be given one grain of rice. At the end of that first week he would return, and asked to be give the double, two grains. On the third visit a week later, again the double, so 4 grains, and so on and so forth for a total of 64 visits, each visit doubling the quantity of rice. Of course the Emperor quickly agreed to this simple request, without knowing that on the last visit he would have to give the inventor 2^{64} grains of rice, so 2 multiplied by itself 64 times, which would be far more rice than existed in the whole of China.

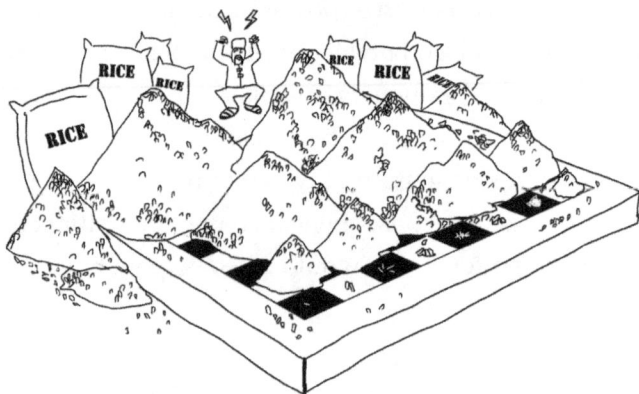

To give an example of a natural complexity, consider an average molecule, say the caffeine molecule, swallowed down daily by billions across the globe. It turns out that to simulate one single caffeine molecule on a standard digital computer, using regular bits of 1's and 0's, we would need a total of 10^{48} bits, which is such an incredibly large number that no single super computer in the world could ever hope to hold this much information.[3] Indeed the number is so big that if every bit of information, every 1 and every 0 were each held on a single atom, then the 10^{48} atoms needed is estimated to be about 10% of all the atoms that go to make up the entire Earth. So just imagine trying to come up with a computer simulation combining caffeine with the next new artificial sweetener or flavourer. Yep, you got it, impossible. Of course this is where quantum can kick in. A quantum computer does not work with conventional bits, those simple 1's and 0's we are so accustomed to, but rather it relies on the appropriately named *qubit*, or 'quantum bit'. And what a qubit represents is a completely different technology that fundamentally changes the way we need to think about, and interact with computers.

Of course qubits work in a quantum way, with quantum characteristics which allow them to handle much more information than a standard bit, and more importantly, handle multiple computations simultaneously. Whilst a classical bit can be in a state of either 1 *or* 0, a qubit can be in a state simultaneously of both 1 *and* 0. This is what we referred to in an earlier chapter as *superposition*, and is the basic property that gives qubits their special computing characteristics. All this sounds a bit odd, and it is, don't worry, such is the non-intuitive world of quantum mechanics. The main point is that in conventional digital computing, when a new bit is added, we are simply adding in a *linear* way one new piece of data capacity, or computing power. When we add qubits, we multiply up *exponentially* the computing power. So like the grains of rice on the chessboard, or the extra guest at the wedding, even small combinations of qubits can have tremendous computing capacity. At 100 qubits the number of entangled states available for computing is so enormously large that it is a number greater than all the atoms that go to make up the Earth. At 280 qubits, the whole universe.[4]

Bits	Possible States	Bits Available
2 Bits	1-1 or 1-0 or 0-1 or 0-0	2
3 Bits	1-1-1 or 1-1-0 or 1-0-1 or 1-0-0 or 0-0-0 or 0-0-1 or 0-1-0 or 0-1-1	3
10 Bits	$2^{10} = 1,024$	10
N Bits	2^N	N
Qubits		
2 Qubits	1-1 and 1-0 and 0-1 and 0-0	4*
3 Qubits	1-1-1 and 1-1-0 and 1-0-1 and 1-0-0 and 0-0-0 and 0-0-1 and 0-1-0 and 0-1-1	8
10 Qubits	$2^{10} = 1,024$	1,024
N Qubits	2^N	2^N

(* a superposition of 4 quantum state can carry simultaneously 4 bits of information, 1 for each state... etc)

Probably the most important characteristic of qubits is that of their ability to handle multiple computations simultaneously. So whilst a conventional computer has to crunch through all the alternatives in sequence to arrive at an answer, the quantum computer can work simultaneously on many possible alternative solutions, homing in and coming up with the required outputs much faster. In October 2019 to a great deal of media fanfare, Google, using a 53 qubit machine, announced that they had achieved 'quantum supremacy', solving an arbitrary mathematical problem in 200 seconds when an ordinary supercomputer would have taken 10,000 years. The claim was immediately contested by their competitor IBM, who claimed that one of their conventional supercomputers could have done the same calculation in two and a half days. However even if IBM were right, you get the basic idea.[3]

Quantum computers are not easy, nor indeed inexpensive to build. In their operating mode the qubits are very delicate creatures, with even slight disturbances likely resulting in a collapse of their quantum states. They need to be kept away from all types of interferences such as vibrations, radiation, light, or any other form of electromagnetic wave. Neither is there an agreed and accepted technical solution for their manufacture. The qubit itself can be any form of gateway that can act in a quantum way, and be connectable to other qubits, so theoretically it can be made from individual particles, atoms, molecules or even micro structures such as those employing superconducting metals. This latter method has been the most popular in the early development of quantum computing, preferred by both Google and IBM. The drawback is that the whole system needs to be encased in liquid helium in order to maintain the extremely low temperatures necessary to achieve superconductivity, i.e. temperatures close to zero degrees Kelvin, or the 'absolute zero' of empty galactic space. A second more recent method of qubit manufacture has been using electrons, or quantum dots, embedded in silicon, with the advantage that the qubits can operate at somewhat higher temperatures and with apparently enhanced stability. Either way, the whole set up remains extremely complicated, and it seems very unlikely that quantum computing will ever be incorporated into our lap tops or mobile phones anytime in the near future.

The most likely role for quantum appears to be some form of 'cloud computing' solution, similar to that currently used by IBM, operative since 2016 with over 230,000 users worldwide. What IBM have set up is an online web interface to their quantum computing centre, where users can access and input their algorithms to be run on their recently upgraded 53 qubit machine. It is not even necessary to interface in any special new quantum programming language, since the idea is that both input and output should be of a standard micro processing nature.[3]

So the immediate future appears to place quantum computing as a sort of high tech solution for carrying out difficult computations that digital computers are incapable of handling. A typical task perfectly geared for quantum would be one with a relatively small number of inputs and outputs, but a vast number of possibilities in between. An example is the travelling salesman problem, otherwise known as route planning. Imagine you had to leave your office O, visit three customers A, B and C, before returning to the office, and the problem is to calculate the shortest route. Well in this example there are only a small number of variables, and so even without a computer you could quickly establish that there are only six alternative routes (which in reality are three, since the other three are identical only in reverse order). It is then a simple task to crunch the calculations and decide on the quickest. Of course increasing the number of customers turns this into an exponential problem. For 5 customers there are 12 possible routes to consider, whilst for 10 there are 181,440. By the time we get to 15 customers there are over 87 billion possibilities, which if our lap top took just one microsecond for each computation, it would be running for more than 12 hours. At 20 customers, it would need nearly 2,000 years.[5]

Real world problems are much more complex, and so it appears that for many, quantum computing is really the only alternative, with possible applications being limitless. We touched on molecular modelling, which could help in any number of fields such as chemistry in the development of new compounds, materials, foods or pesticides, or in medicine in the research of new drugs, treatments, biotech applications, pandemic planning etc. Quantum can help with our forecasts of weather systems, simulations of global warming, running economic models or optimising investment portfolios. And for sure it is a no brainer in the field of encryption and cybersecurity. Even since the early 1990's it's been known that quantum computers would have little problem cracking many of the protection systems currently used around the world in on-line security, with of course the flip side being their necessity in creating completely new quantum security solutions.[6] Finally quantum computing will be a key ingredient in our development of artificial intelligence, or AI it is more popularly referred to.

At exactly what point we will be able to claim to have achieved true AI is probably a question as much for philosophers as for physicists. Whatever the definition, we are getting closer all the time. After Big Blue's chess victory in 1996, another milestone was reached in 2017 when Google's Deep Mind machine claimed the scalp of the world champion of the Chinese game Go. This ancient board game is of a complexity far superior to that of chess, with the board consisting of 19 x 19 squares compared to the humble 8 x 8 of chess. To achieve this feat the operating system incorporated 'machine learning', whereby following a simple set of initial parameters, combined with a 'carrot and stick' reward system, the computer basically taught itself to play.

Following developments like this it becomes easy to imagine that if we can successfully bundle together the brute force of traditional computers with machine learning, language and voice recognition, universal data mining, a few other useful skills, and combine the whole lot with access to quantum solutions, then, well, it will certainly look very much like AI in most people's book.

At the beginning of this chapter we contrasted digital and quantum computing using a candle - light bulb analogy, which was perhaps unfair in the sense that quantum looks extremely unlikely to ever be a replacement technology. Traditional microcomputers are robust, fast, reliable and depending on the task, offer a much more efficient solution than quantum. We will never need quantum to run Windows, nor watch a film. What quantum can bring to the party is its unique capacity for handling complex problems of an exponential nature, taking us into areas where it would simply be impossible using traditional computing. With this in mind it is perhaps better to illustrate the two as a likely couple, where the sheer brute force, number crunching efficiency of our microprocessor complements wonderfully the very delicate, multitasking nature of quantum. Put them together, and it will be a formidable marriage indeed.

At the Nano Bottom

"There's plenty of room at the bottom": Richard Feynman

Again Feynman, December 1959, Caltech, gave a now celebrated address to the American Physical Society entitled 'There's Plenty of Room at the Bottom', where he explored theoretically the futuristic possibilities of design and engineering on the nanoscale. The emphasis was on 'plenty' as Feynman set out to demonstrate the *"staggeringly small world"* that we have below us.

Of course we are completely accustomed to hearing about things on the biggest scale, the millions of light years from here to the next galaxy, or the gazillions to the edge of the universe. On this scale of course we are observing from the bottom-most rung, and our perspective is on our own tiny insignificance. However if we start to look the other way, we can see that actually, in the grand scheme of things, we can consider ourselves to be somewhere near the middle when it comes to relative distances and sizes.

To go up the size scale we just add zeros. So to our trusty one metre, we just add 4 zeros to get the 10,000m, or 10km, we have to cover to drive to the next town. However if we wanted to fly to the Moon, then we would have to cover the distance of a few meters followed by 8 zeros, and to the Sun, 11 zeros. To get to the nearest star, Proxima Centauri, we would need to add 16 zeros, and to the nearest galaxy Andromeda, we would need to scale things up to 22 zeros. To cross the whole observable universe, well, 26 zeros would just about do it.

What about the other way? To go down in size we also need to add zeros, but this time in front of the meter. So with 3 zeros and our decimal point we would be at 0,001m, or 1mm, just a bit larger than a grain of sand. With 5 zeros we get to around the width of a human hair, and already starting to get towards the limit of the observable to the human eye. At 9 zeros we would be down at the 'nanometre' scale, where we can find molecular structures like DNA which contains the blueprint of all life on Earth, as well as Covid-19 and other viruses which so torment our life on Earth! At 12 zeros we enter the world of atoms, from caesium the largest at 265 'picometers' to helium the smallest at 31pm. At 15 zeros we finally get to the size of the atomic nuclei and the individual protons and neutrons that go to make them. However even at these tiny levels, we would still be very, very far from the smallest scale considered in quantum mechanics, the Planck length, which with its 34 zeros is firmly at the bottom of the list.[1]

Nanotechnology is about building and manipulating things on a very small scale, on the nanoscale. Nano is extremely small. If you took something that measured 1nm, tiny 'nano-people' say, and lined up 10 million of them shoulder to shoulder, so something like the population of Sweden, the line would stretch no further than 1 centimetre across your fingernail. Or if you are a guy and you had a wet shave this morning, dragging your razor from left cheek to right, in the couple of seconds it took you to complete the stroke, the recently cut whiskers on the left cheek would have grown by approximately 10nm.[2] This is the world we are in when we talk about nanotechnology, and we are already there, building structures and manipulating materials on this tiny nanoscale.

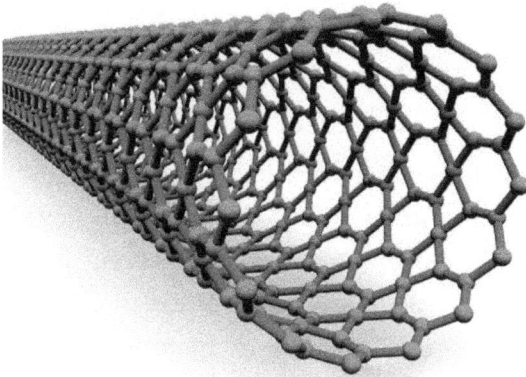

Carbon nanotube 2nm. Its excellent strength to weigh ratio has found uses in fields as varied as aerospace and sports equipment.

In his 1959 address[3], the basic ideas put forward by Feynman have been almost a road map of the route we have followed during the intervening years. For example he pondered on the benefits of miniaturizing the computer, which in those days were enormous, filling whole rooms. He played with the *"wild idea"* of being able to *"swallow the surgeon"* and sending some very tiny machine into the blood vessels and get it into the heart to have a *"look around"*. Of course he envisaged things at the very smallest level of the atom, where the effects of quantum mechanics come into play, and could be made use of. Importantly he talked about the possibilities if one day we were able to control and manipulate the very atoms that go to make up material, commenting:

"What could we do with layered structures with just the right layers? What would the properties of materials be if we could really arrange the atoms the way we want them? They would be very interesting to investigate theoretically. I can't see exactly what would happen, but I can hardly doubt that when we have some control of the arrangement of things on a small scale we will get an enormously greater range of possible properties that substances can have, and of different things that we can do."

In a nutshell this is the very heart of nanotechnology, being able to manipulate at an atomic and molecular level in order to build new structures and materials that not only have interesting possibilities due to their tiny size, but can also have completely different properties compared to materials and compounds occurring naturally. Just imagine being able to construct a material with the transparency of glass, the conductivity of copper, and the flexibility and durability of plastic all rolled into one. Or some new compound for manufacturing more efficient, bigger capacity, longer lasting, non-polluting batteries. Of course manipulating atoms and molecules has been what chemists have been doing for centuries, and there is indeed an overlap. However the difference that Feynman was driving at is subtle, where instead of creating circumstances so that atoms can bond naturally (i.e. throwing everything into a test tube and giving it a good shake), he envisaged being able to work with individual atoms, one by one, like a child with Lego. Well, for practical human purposes, atoms are about as small as it gets. A cube with sides of 1nm can easily contain a few thousand atoms, and being able to work with Lego bricks as small as these is at the very forefront of nanotechnology.

One of the first publicised atomic manipulations was in 1989, when IBM succeeded in arranging 35 individual xenon atoms onto a chilled crystal of nickel to form their famous company logo. The scanning tunnelling microscope technology developed by their in-house engineers Gerd Binning and Heinrich Rohrer to achieve this earned them a Nobel Prize, and was the same technology used again in 2013 to this time earn the company the Guinness World Records record for the world's smallest stop-motion film 'A Boy And His Atom'.[4] (Check it out on YouTube)

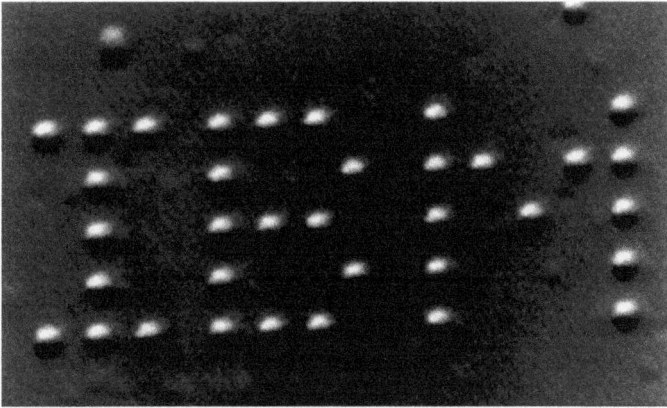

Due to developments since the early 1960's, we are of course extremely familiar with the continual downsizing of both the computer, and the electronic industry in general. This is not really a surprise, since it has been in these sectors where most of the investment has gone, and earlier paybacks achieved. However as we move into the 21st century, nano applications are being investigated and successfully applied across an extremely wide variety of sectors, with developments ironically being helped swiftly along by all that computing power now available to us.

In Europe, one of the leading R&D centres for nanotechnologies is the MESA+ institute at the University of Twente, established in 1981 to "harness the properties of materials manipulated on an atomic or molecular level".[5] The work going on at MESA+ is typical of such institutions and is extremely wide ranging, covering fields as diverse as energy, health, security, food and water, materials and coatings, artificial intelligence and robotics amongst many others. One of the specialties at MESA+ has been the development of new materials, where they have pioneered techniques using high energy laser pulses for building multiple atom-thin layered structures with specific properties not found in naturally occurring materials, with possible uses in new electronic and optical applications. Another area of much research has been that of health and medicine, where nanotechnologies have been developed to help with disease detection, drug development and even treatment simulation using the aptly named 'organ-on-a-chip' technologies. An extension of this latter idea is the 'brain-on-a-chip', which may eventually find uses in other areas such as computing or energy saving electronic devices. Another project, unusually close to the "*wild idea*" of Feynman, involves a 'nano-pill' for the early diagnosis of colon cancer, which consists of a small contraption almost resembling a mini submarine, with on board electronic sensors, camera, a 'body heat' power system, and even an embedded antenna for transmitting real time results to the patients mobile phone.[2]

Of all the 'new technologies', nanotechnologies are without doubt the most versatile and the most effective in finding new solutions to real world problems, and look set to be at the scientific leading edge for a long time to come. A confirmation that good things do indeed come in small packages!

The Dark Universe

"We know that God is everywhere; but certainly we feel His presence most when His works are on the grandest scale spread before us; and it is in the unclouded night-sky, where His worlds wheel their silent course, that we read clearest His infinitude, His omnipotence, His omnipresence": Charlotte Brontë 'Jane Eyre'

For as long as human beings have walked the Earth, our eyes have been drawn upwards to gaze at the vast multitude of celestial bodies that fill the night sky, in search of scientific understanding, spiritual guidance or simply to wonder in awe at the majesty of the heavens. Not surprisingly, we know an awful lot about the universe. It seems however that there is even more we do not know.

We know that the universe is 13.8 billion years old.[1] By comparison our Earth was formed about 4.5 billion years ago, and Homo sapiens have been kicking around on it, well, let's say a couple of hundred thousand years. Complex societies with city states, writing, coinage and the rest, have sprung up only in the last 6,000 years.[2] So if the life span of the universe was squashed into a 24 hour clock, then the whole of recorded human history would be a brief flash of a few insignificant fractions of a second just before midnight. And within that tiny sliver of time, it has only been in the last 100 years that we have started to observe and understand the goings on beyond the immediate vicinity of our own solar system and the Milky Way surrounding us. It's really quite amazing that we know as much as we do.

We know that the universe started in a Big Bang. The picture that usually springs to mind is one of some huge atomic bomb like explosion, playing out in our imagination as we watch it from a distance. However this view is wrong in the sense that the explosion didn't take place 'in' anything, and there was no 'outside' from where to observe. In that one brief moment, space itself was created, so the only place we could be if we tried to imagine events would on the inside, visualising things from within. To take it all in however we would need to be paying very close attention. Within the first millionth of a second the space created was already many billions of kilometres across, mega hot, and swimming with a plasma of elementary particles like electrons and quarks. Almost as quickly, the quarks started forming up into protons and neutrons, which in turn fused together into the atomic nuclei of hydrogen and helium.[3] After only three minutes, or the time it takes to make a cup of tea, the universe already contained basically all of the matter that would go on to form everything that surrounds us today. Indeed if we consider that the human body is mostly water, with the most abundant element in us being hydrogen, then it's fun to think that the majority of atoms in our body, or at least their nuclei, were formed 13.8 billion years ago in these first three minutes.

Following these first few explosive moments, we could kick back a bit, since it would take another 380,000 years of expansion for things to cool down sufficiently for the electrons to finally join up with the hydrogen and helium nuclei to form the stable atoms of matter we know and love. 4 Around this time we would witness something else quite significant, that finally things were cooled down and spaced out enough for energy in the form of photons to start circulating. The universe had finally become, in effect, transparent. However it was still a very dark place, and we would need to hang around another 400 million years for gravity to pull together sufficient material to form the first stars and galaxies to finally get a bit of light to cheer things up. According to the book of Genesis, the light trick happened on the first day. According to physicists, it took a little bit longer.

Unusually, the term Big Bang was coined somewhat derogatively in a 1952 radio broadcast by the well-known British astronomer Fred Hoyle, who thought that the theory of everything being squashed up at a single point, followed by a sudden and dramatic energetic expansion, was simply a load of old tosh. The idea of the 'primeval atom' as it was previously referred to had been around by since the 1920's, championed by a Belgian Catholic priest-scholar Georges Lemaître, and given credence through the observations of the American astronomer Edwin Hubble. Hubble was the first astronomer to detect the existence of galaxies beyond the Milky Way, and what he also saw was that these newly discovered galaxies were not only at huge distances from our own galaxy, but that they were all flying away from us, and from each other, at great speeds and in all directions. Of course from that point it was not difficult to extrapolate things backwards, to a time when perhaps all the material had been much closer together, and so the primeval atom theory fitted nicely. Over the years the theory gradually gathered force, and was even cited in 1951 by Pope Pius XII as being a scientific validation of Creationism, although thanks to the subtle intervention of Lemaître, the Vatican was discretely persuaded to resist with proclamations on cosmology, and to sensibly follow a generally more neutral line of 'neither connection nor contradiction' between religious and scientific theory.

The argument was finally settled in the early 1960's by two American astronomers, Arno Penzias and Robert Wilson, working on a Bell Laboratories radio antenna in Holmdel, New Jersey, when they discovered almost by accident the famous Cosmic Microwave Background, or CMB for short.

Cosmologists had speculated that if the Big Bang model were correct, then the massive outpouring of energy in the form of photons just at the point when the universe became transparent would still be around today, only with much less energy, and with photon wavelengths stretched to those of microwaves. For their part, Penzias and Wilson were working on radio astronomy and satellite communications projects, very sensitive work for which their instruments needed to be completely free from any type of interference. For more than a year they beavered away setting things up, but try as they might, they were never quite able to eliminate all interference and rid themselves of what we know commonly as static. Eventually their efforts came to the attention of a group of Princeton University scientists led by Robert Dickie who happened to be working on the very thing that Penzias and Wilson were trying to get rid of, and they recognised that what the pair were picking up was indeed the CMB from the Big Bang. Despite never having set out to find it, nor indeed even interpreting the results, Penzias and Wilson were awarded the 1978 Nobel Prize, with Dickie and his team a mention in the footnotes.[5]

The CMB, as well as giving us proof of the Big Bang, provides us with the earliest photograph we have, or indeed ever will have, of the universe just at the point where it became transparent. As technologies have improved, we have been able to obtain ever more detailed images, such as those provided by NASA's WMAP satellite, or the European Space Agency's Planck Surveyor. As the earliest clue that we have about the state of things, the CMB remains even today the source of much investigation, with ever more detailed images being scoured for hints of gravitons, suggestions of multi-verses, and proofs of other strange and leading edge theories.

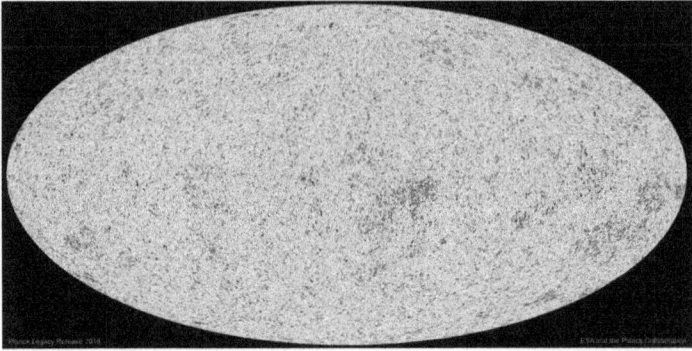

ESA Planck surveyor image of the early universe

We know that the universe is full of stuff. In December 1995, the Hubble Space Telescope was given an unusual mission. Over a period of 10 days it was to take long exposure images of a tiny piece of the sky, roughly the size of a tennis ball 100m away, and where, as far as we knew, there was absolutely nothing at all to see. What emerged, was that not only was there stuff there, but there really was a LOT of stuff there, around 3,000 galaxies, each made up like our own Milky Way of a few hundred billion stars.[6] The Hubble Deep Field image as it is known has been repeated various times since, focusing on other 'uninteresting' parts of the sky, and always with the same result, giving us estimates today of a universe containing as many as 2 trillion galaxies.[7] All of this observable stuff we refer to as ordinary matter, in the sense that it behaves exactly like matter here on Earth, following exactly the same laws of physics. However, despite the unimaginably large amounts, physicists believe that the whole lot represents hardly one sixth of the total amount of matter that actually exists in the universe, and what the rest is, we have little clue. We call it *dark matter*.[6]

Despite its name, and the fact that none has ever been detected, dark matter is not science fiction. There is just too much evidence for it being out there for it to be otherwise. The first clue came after years of studying the behaviour of galaxies, all of which are spinning around some central point, often a black hole. The spinning is a result of the effects of gravity, with the greater the gravity, the faster the spin. Our own Sun for example is traveling through space at close to 200 km per second, or 700,000 km per hour, carrying with it the Earth and all the other planets. Even so, the distances across the Milky Way are so great that it takes about 230 million years for the Sun to complete one full rotation. However, knowing how fast a galaxy is spinning, physicists can calculate how much gravity there is, and hence how much mass there must be in the galaxy. The problem comes when they crunch the numbers to add up all the masses of all the stars and all the dust and other material visible in the galaxy, and the total always comes out short, way short, like 'six-times-less' short, and this result seems to be consistent across all galaxies studied. The conclusion is that in each galaxy there is a huge amount of mass floating about basically invisible to us.

A second clue comes from the way galaxies bend light, as predicted by Einstein and first measured by Eddington and Dyson in 1919. As light passes nearby galaxies, it is bent due to the gravitational forces, with the greater the gravity, the more it is bent. Again the observations imply that the mass of the intervening galaxies appear six times greater than that calculated simply by summing up the masses of all the observable material. A final pointer has come in recent years via the advent of computer modelling to simulate galaxy and star formations in the early universe. It turns out that without the presence of something akin to dark matter, there would be insufficient gravity for galaxies and stars to form simply on their own.

So there appears to be a huge amount of dark matter out there, matter that does not emit or absorb light, or any other form of radiation, does not interact with itself, nor with ordinary matter, nor react to any of the normal forces, other than gravity. And if it does exist, where does it come from? Was it formed in the Big Bang? What laws of physics does it obey? Since the idea of dark matter reared its head, lots of smart theories have been proposed, although all are quite literally, shots in the dark. The most popular idea supposes that dark matter is made up of particles, with properties completely different to any of the ordinary matter that forms our Standard Model. For the numbers to add up, these supposed dark matter particles must be everywhere, with perhaps a million or more whizzing through your room right now, at speeds of around 100 km per second, passing straight through you, the floor, the Earth, and out the other side, or back the other way, without interacting with anything 'en route'. To do this they must be unbelievably small, with the million or so flying about your room estimated to weigh in total no more than a millionth of a millionth of a millionth of a kg.[6] So how to detect something that small when it refuses to interact with anything? As always physicists are not short of ideas. Astronomy and the study of galaxies is one possible source, as are particle accelerators like the LHC at CERN. Another is using underground detectors similar to those used to detect neutrinos, a number of which are currently in operation around the world such as the Xenon detector of the Gran Sasso Institute deep below the Apennine Mountains in central Italy.

An alternative idea to explain the observations related to dark matter that has been gaining traction over the last few years postulates that they are not caused by the presence of mysterious material at all, but rather due to some fundamental anomaly in the workings of gravity itself. Were this the case, it would require some deep rewriting of Einstein's theory of general relativity, certainly not a trifling matter. Either way, the search for dark matter, and the possibility of discovering some 'new physics' is one of the most important areas of research today, although for the time being at least, we remain, as they say, in the dark.

We know that the universe is big, really big, and that it is getting bigger. The universe is so big we have to measure the distance in *light years*. Light travels fast. It takes just 1,2 seconds to make the 385,000 km to the Moon, a distance that the Apollo astronauts took 3 days to cover. So just try and imagine how far that same beam of light would travel in a whole year. Answer: 9 trillion km's, that's 9 followed by 12 zeros, or 1 light year. It seems huge, but on the galactic scale, it's insignificant. The distance to the nearest galaxy, our galactic neighbour Andromeda, is a mere 2.5 million light years, so to then contemplate distances on the scale of the whole universe, well, we're talking lots of zeros. The best estimate we have today of the size of the observable universe is around 90 billion light years across,[8] even though it may well be that the actual universe is much bigger still, with any light or other form of radiation from further out yet to cross the immensity of space and reach us here on our lonely planet.

We know that the universe is expanding, with all the galaxies flying away from each other as Hubble first discovered back in the 1920's. However the important point is that the stuff is not flying 'through' anything, rather it is space itself that is expanding. The image often evoked is that of a fruit cake, puffing up as it cooks away in the oven, and carrying along with it all the pieces of fruit. Similarly it is the whole volume of space that is increasing, with new space being created all the time, and the universes getting bigger in the process. The other thing we know, based on observations first made in the 1990's, is that the *rate* at which space is expanding is also increasing, or accelerating, meaning that the amount of 'new space' that will be created in the next billion years is more than the amount that was created in the last billion, and so on, faster and faster. So unlike the fruit cake in the oven which grows quickly at the beginning and then slows to its final size, the universe according to today's most accepted theories is simply getting bigger and bigger at a runaway accelerated rate. And if it is growing in this way, then something must be driving it, some force or some energy. Again we have absolutely no idea what this energy could be, and the name physicists have given it is? You guessed correctly, *dark energy*.

The basic idea behind dark energy is a curious one. Imagine a box containing some ordinary energy in the form of say some particles, or some photons of light or other radiation. Double the volume of the box and the ordinary energy would remain the same, just now spread out through a volume twice the size, or to half of the original density. Dark energy on the other hand, as we doubled the size of the box, would also double, and the density would remain constant. The more we increased the volume of the box, the more dark energy we would be adding.[3] This concept, apart from being curious, appears to contravene accepted fundamental laws of physics, such as how energy can neither be created nor destroyed. Just as with dark matter, it appears that we are dealing with a completely different animal that is beyond our current understandings. What's more, doing some back of the envelope calculations, the observations imply that the amount of dark energy out there is huge, with more and more being created all the time. From special relativity we know that energy and mass are simply manifestations of the same thing, so if we take a snapshot of the universe today, it turns out that it is made up of approximately 70% dark energy, 25% dark matter, with only a tiny 5% being the ordinary matter that we know and love so well. Put the other way around, an incredible 95% of everything in the universe is stuff about which we have absolutely no idea at all. At least as far as our understanding goes, the universe we live in is a very dark place indeed.

So what of the future? Well, if current theory is correct, the picture is not a great one, although there is a very good chance that humans will not be around to witness very much of it. After nearly 5 billion years, our own Sun is already about half way through its life cycle, after which it appears destined to swell up to mega size, destroying the Earth in the process, before exploding off its outer layers and shrinking back on itself to end its days as a tiny, slowly cooling white dwarf star. In the meantime the universe will continue to expand, with all the energy and all the matter, dark or otherwise, getting stretched out over ever greater distances. Stars and galaxies will eventually burn themselves up, and there will come a point when the remaining material will be so far apart, so stretched out, that it will be unable to reform into new ones. Eventually a good part of this material seems destined to end up in mega black holes, which themselves will gradually disappear as they very slowly radiate themselves away in the form of low energy Hawking radiation, so named after its discoverer Stephen Hawking. The universe will then have completed its cycle from Big Bang to Heat Death: from a smooth, uniform, very high density, very high temperature singularity of energy, to a smooth, uniform, very low density, very low temperature expanse of dead space, populated only by low energy photons and a sparse distribution of elementary particles.[4&7] And it is fitting perhaps that the final act in this incredible story be scripted in a roundabout way by Hawking himself, perhaps finally concluding his search for the theory of everything, where it appears that the eventual 'everything' will be little more than eternal nothingness.

So is that really it? Is our universe really doomed to slowly peter out into a vast expanse of nothing and us along with it? Well, according to current theories it certainly seems to be the case, although ever the optimists, there are perhaps a few pegs where we can hang our optimists hat. Donald Rumsfeld, United States Defence Secretary under President George W. Bush once made a famous remark about the *known unknowns* and the *unknown unknowns*. As we have seen throughout this book, out there in our universe we have some very significant known unknowns, which if we are ever able to decipher will undoubtedly throw up new physics, new theories, new technologies and new reads on the future of the cosmos. In turn, this knowledge will reveal to us the unknown unknowns, the future frontiers of exploration, challenging us to build our understandings to even greater heights. And for all this, time is on our side. The complete estimated lifespan of the universe is on a time scale so fantastically huge, so many gazillions of years that the sheer immensity escapes even our fertile imaginations. Considering that the knowledge we have today is based on little more than a brief 100 years of perseverance, and reflecting on how far we have travelled in this insignificant fraction of time, then perhaps we should be rather hopeful about the long and winding road before us, where it may lead, and what we may learn along the way.

And so, in this vein it would be apt to give the final word to a physicist, to John Archibald Wheeler, the man who popularized the term 'black hole', and one of the most influential American physicists of the 20th century:

> *"We live on an island surrounded by a sea of ignorance. As our island of knowledge grows, so does the shore of our ignorance."*
>
> *John Archibald Wheeler*

Acknowledgements

To Diego Ortola, for his time, persistence and creative illustrations. To my good friend Thomas Lovenskiold, the smartest guy in Oslo, for his fact checking and error spotting. Finally to Jouly, for her understanding, patience and moral support, without which this humble book would never have gotten out of the project shed.

About the author

Originally from North Wales, Gary Lewis studied at Dinas Brân Comprehensive School in Llangollen, Denbighshire, before earning his physics degree at the University of Durham in 1986. In 1991 he moved to Spain where he completed an M.B.A. at the IESE Business School. His professional career has been dedicated to sales and marketing in the fmcg & fashion sectors. He has two children, muddles about in five languages, and currently lives near Barcelona.

Notes

Foreword

1. Richard Feynman 'The Complete FUN TO IMAGINE with Richard Feynman BBC2' (You Tube / 1 Nov.2018) *https:// youtu.be/ P1ww1IXRfTA*
2. Tech Planet 'Top 7 Emerging Technologies That Will Change Our World!' (You Tube / 17 Apr.2019) *https://youtu.be/ hTe2PYwnEpcTxRldL2CDBk*

The Theory of Everything

1. David Tong 'Quantum Fields: The Real Building Blocks of the Universe' / Royal Institute (You Tube / 15 Feb.2017) *https://youtu.be/zNVQfWC_evg*
2. Sean Carrol 'The Big Picture' / Royal Institute (You Tube / 22 Nov.2017) *https://youtu.be/2JsKwyRFiYY*

'Annus Mirabilis' 1905

1. Democritus is generally credited with creating the 'atomist' school of thought, which considered all substances made up of small indivisible entities which they termed 'atoms',

a Greek word meaning 'uncuttable' or 'indivisible'. The opposing opinions of the time, held by such figures as Aristotle, believed everything in the world to be made up of combinations of the four basic elements: earth, fire, air and water plus an invisible aether.

2. 'On The Electrodynamics of Moving Bodies' can be read on-line at einsteinpapers.press.princeton.edu. See also Bill Bryson 'A Short History of Nearly Everything' Black Swan 2004 / p.160

3. Isaac Newton English mathematician, physicist, astronomer, theologian and author of 'Mathematical Principals of Natural Philosophy' first published in 1687 which laid the foundations of classical mechanics.

4. The velocity of light in a vacuum 'c' measured by Michelson and Morley just under 300,000 km per second was in agreement with the electromagnetic equations of the Scottish mathematician and scientist James Clerk Maxwell.

5. Russel Stannard 'Relativity A Very Short Introduction' Oxford University Press / Revised impression 2017/ p.9

$E=Mc^2$

1. Russel Stannard 'Relativity A Very Short Introduction' Oxford University Press / Revised impression 2017/ p.37 p.39 p.40

2. 10.4 megatons is equivalent to $4,3 \times 10^{16}$ Joules of energy, which, by Einstein's equation, divided by c^2 give us approximately 0,5kg

3. A normal hydrogen atom has one lone proton in its nucleus. It is the first, and most basic element in the periodic table. Deuterium is also a hydrogen atom, but with one neutrally

charged neutron added to the nucleus, and is 'heavy' since due to the added neutron it has almost twice the weight of normal hydrogen.

4. Additional elements to be used as fuel would be tritium, an even heavier hydrogen atom with two neutrons in the nucleus, and lithium, the third element along in the periodic table. Whilst lithium is naturally occurring and available in reasonable quantities, for example in brine, tritium is scarce, but could be bred within the reactor itself.

5. Iter.org

6. Bill Bryson 'A Short History of Nearly Everything' Black Swan 2004 / p.161-162

Space-Time

1. Arvin Ash 'General Relativity Explained Simply' (You Tube / 20 Jun. 2020) *https://youtu.be/tzQC3uYL67U* & Spark 'How Gravity Affects Time' (You Tube / 4 Jan. 2019) *https://youtu.be/QQRj78jOxWo*

2. Commander Jim Lovell to mission control April 13th,1970_ *https://www.nasa.gov/mission_pages/apollo/missions/apollo13.html*. Ref also BBC World Service Podcast '13 Minutes to the Moon' Season 2 / Apollo 13

3. As the Moon orbits the Earth, it slowly rotates in such a way that from the Earth we only ever see one side, always the same side, of the Moon. The first humans to set eyes on the 'dark side' were the crew of Apollo 8.

4. The point where the gravity pull of the Moon is exactly equal to the pull from the Earth is known as a Lagrange point, some 320,000 kms from Earth and 60,000 kms from the Moon. It is like the top of a rounded hill, from where the smallest

push would send a stationery object 'rolling' down one side or the other.

5. GPS Global Positioning System is just one such system, one of the first and most well-known. Other similar systems exist, and the generic term for all is GNSS Global Navigation Satellite Systems. Ref. David Barry 'Sextant: A Voyage Guided by the Stars and the Men Who Mapped the World's Oceans'

6. 50 nanoseconds is the time it takes the electromagnetic signal from the satellite to travel 15m, the desired accuracy.

7. Clifford M.Will the James S.McDonnell Professor of Physics at Washington University, St Louis, author 'Was Einstein Right'

8. When a star gets towards the end of its active life, and the outward pressure from fuel combustion is not sufficient to balance the inward gravitational forces on the burnt material, then the whole lot can collapses in on itself with exotic results. The most extreme is the famous 'black hole', so called because it creates an area of space so warped and bent that not even light can escape the extreme gravity.

9. Laser Interferometer Gravitational-Wave Observatory.

Waves are Particles

1. Dr Richard Gunderman 'Tesla The Man The Inventor And The Age Of Electricity' Andre Deutsch 2019

2. Quanta, plural of quantum, from the Latin 'quantus' meaning 'how great'

3. Planck's constant h is one of the most fundamental numbers in our universe, right up there on the podium alongside

others constants such as the velocity of light *c*, or our geometrical friend *pi*.

4. The Planck length is a 'naturally occurring' constant in the sense that it is derived directly from three fundamental constants of our universe: the speed of light *c*, Newton's gravitational constant G, and Planck's constant *h*. Similarly physicists have derived the concepts of Planck time and Planck energy, important for the development of theories such as Loop Quantum Gravity. Ref also Arvin Ash 'Visualizing the Planck Length. Why is it the Smallest Length in the Universe' (You Tube / 12 Oct. 2019) *https://youtu.be/ bjVfL8uNkUk*

5. Carlo Rovelli: 'Reality Is Not What It Seems' – The Journey to Quantum Gravity & 'Seven Brief Lessons on Physics' Penguin / Grains of Space

6. The photoelectric effect is popularly used in light sensors and security systems, whereby a light beam onto a metal surface creates a small amount of electricity and so completing an electric circuit. If the light beam is broken, even momentarily, so is the electric current, and the resulting interruption can then trigger an alarm, open a door, or some other similar such mechanism.

7. Einstein calculated the energy E of each photon was related to the frequency *v* of the wave via the Planck constant *h*, giving yet another famous equation E = *hv*.

8. Bill Bryson 'A Short History of Nearly Everything' Black Swan 2004 / p.183

9. The first physicist to suggest and depict this type of structure was a Japanese scientist called Hantaro Nagaoka, who envisaged an orbiting electrons structure, rather like

Saturn and its rings. The Nagaoka model was mentioned by Rutherford in his 1911 paper.

10. The model of the atom was completed in 1932 by the discovery of the neutrally charged neutron by James Chadwick, at the Cavendish laboratory, Cambridge, which helped explain how the nucleus could be held together with so many positively charged protons so close to each other.

11. In 1965 became the Niels Bohr Institute.

12. 'Probably the best lager in the world' 1970's Advertising Campaign.

Particles are Waves

1. John Polkinghorn 'Quantum Theory A Vary Short Introduction' Oxford University Press 2002 / p.19

2. Of the TV show Big Bang Theory for non-fans. The Big Bang Theory TBBT S02E02 'Loop Quantum Gravity Vs String Theory' *https://youtu.be/FMSmJCKaaC0*

3. An analogy for Heisenberg's proposal would be a clock pendulum, which in the classical world, when stopped, we would know with certainty both its position (at the bottom) and its speed (zero). However according to Heisenberg, on the tiny quantum scale there would always be some uncertainty, as if the small wave-like quanta making up the pendulum were always 'jiggling about' and we could never quite pin the whole thing down.

4. A few years following his death in 1984, a plaque in honour of Dirac was unveiled at Westminster Abbey inscribed with his famous equation.

5. The use of a probabilistic wave function to describe a quantum particle is probably the most counterintuitive

aspect of quantum mechanics, the basic difficulty being that whilst the particle is in the 'unseen' state we unfortunately have no real world way of describing it neither in terms of images nor language. Think about making a drawing of a person running, easy. Now think about making a drawing of a person NOT running. Difficult. The only thing we can do is to draw again a person running and drop over it some abstract symbol that hopefully helps to conjure up the image we require. Our 'in-between' particle state is like the not running person, with the only way to view it through some abstract mathematical formula.

6. The positron was discovered in 1932 by Carl Davis Anderson of Caltech, who subsequently received a Nobel Prize for his work in 1936. The antiproton was discovered in 1956 by the US physicists Chamberlain and Segrè of Berkeley who were in turn awarded the Nobel in 1959.

7. Tara Shears 'Antimatter : Why the anti-world matters' / Royal Institution (You Tube 18 Oct 2013) *https://youtu.be/ 0Fy6oilRwJc*

8. Dustin Cavanaugh 'Human Teleportation? Quantum Entanglement' (You Tube 18 Nov. 2016) *https://youtu.be/ hTe2PYwnEpc*

Building Blocks

1. BBC World podcast series 'The Bomb' / www.bbc.co.uk

2. Website *home.cern*. The WWW at CERN was developed by Tim Berners-Lee, a British scientist. It was put in the public domain on 30th April 1993, with a subsequent release making

available an open license to maximise its dissemination. The rest, as they say, is history.

3. Bill Bryson 'A Short History of Nearly Everything' Black Swan 2004 / p.211

4. Discovered in an experiment set up by Clyde Cowan and Frederick Reines. Reines was eventually to receive a Nobel Prize for this work, many years later, in 1995.

5. Heavy water contains a relative abundance of deuterium, water molecules H^2O with a neutron in the nucleus of the hydrogen atom.

6. After the Scottish physicist Peter Higgs, one of the proponents of this new field mechanism.

7. Check out the excellent documentary Particle Fever (2013) which at time of writing was freely available on You Tube / 16 Apr.2019 *https://youtu.be/ akCJc/K3DUU*

8. Periodic table of elements in its modern format credited to the Russian chemist, Dimitri Mendeleev first published 1869

Fields that Form the World

1. Richard Feynman 'The Complete FUN TO IMAGINE with Richard Feynman BBC2' (You Tube / 1 Nov.2018) *https:// youtu.be/ P1ww1IXRfTA*

2. Richard Feynman, together with colleagues Schwinger and Tomonaga received the Nobel Prize for their contributions in 1965. Dirac had received his already in 1933 alongside Schrödinger for their initial work on quantum mechanics.

3. David Tong 'Quantum Fields: The Real Building Blocks of the Universe' / Royal Institute (You Tube / 15 Feb.2017) *https://youtu.be/zNVQfWC_evg*

4. We would actually have to create and electron-positron pair, $E = 2mc^2$ where m is the electron rest mass, so 1,022MeV
5. Carlo Rovelli : 'Reality Is Not What It Seems' – The Journey to Quantum Gravity & 'Seven Brief Lessons on Physics' Penguin / Grains of Space
6. The supposed graviton is the name given to the particle of the gravitational field, should the gravitational field be of a quantum nature. In this sense the graviton is to the gravity field what the Higgs boson is to the Higgs field.
7. PBS Space Time 'What are the Strings in String Theory' (You Tube / 18 Oct. 2018) *https://youtu.be/ k6TWO-ESC6A*
8. The Big Bang Theory TBBT S02E02 'Loop Quantum Gravity Vs String Theory' *https://youtu.be/FMSmJCKaaC0*

Universal Measurements

1. Ken Alder 'The Measure of All Things' / Abacus 2004 / Prologue & p.91 p.142 p.145 p.147 p.265 p.364
2. The English Magna Carta of 1215 contained the lofty promise of unified weights and measurements, backed up by parliamentary decrees, and reaffirmed in the Article of Union between England and Scotland of 1706. Despite this, popular practices remained, and the British system was no less complex than the French.
3. The eventual definition of the gram of 1799 was established as the weight of one cubic centimetre of rain water at its temperature of maximum density of 4 degrees Celsius
4. Since 1967 the *second s* is defined in terms of the vibrational frequency of the caesium atom, specifically: the duration of 9,192,631,770 periods of the radiation corresponding to the transition between the two hyperfine levels of the ground

state of a caesium 133 atom. An atomic clock based on the caesium atom such as the caesium fountain at the NPL in the UK (npl.co.uk) is accurate to 1 second in 180 million years. *Ref. BBC World Service bbc.co.uk podcast ELEMENTS / caesium*

Quantum Computing

1. Richard Feynman 'Computer Heuristics Lecture' (You Tube / 3 Jun 2012) *https://youtu.be/EKWGGDXe5MA*
2. Stephen Fry 'Great Leap Years' Podcasts - Season 1 Episode 6
3. ColdFusion 'Quantum Computers Fully Explained' (You Tube / 27 May 2019) *https://youtu.be/ PzL-oXxNGVM* & CNBC 'The Hype Over Quantum Computers, Explained' (You Tube / 10 Jan. 2020) *https://youtu.be/ u1XXjWr5frE*
4. Dario Gil / IBM 'The Future of Quantum Computing' (You Tube / 8 May 2020) *https://youtu.be/ zOGNoDO7mcU* - 53 qubit launched October, 2019.
5. Forbes 29th May, 2020. Routs: OABCO / OACBO / OBACO / OBCAO / OCABO / OCBAO
6. Shor's algorithm, a theoretical way using quantum to break security codes relying on the factorization of large numbers into primes.

At the Nano Bottom

1. Arvin Ash 'Visualizing the Planck Length. Why is it the Smallest Length in the Universe' (You Tube / 12 Oct. 2019) *https://youtu.be/ bjVfL8uNkUk*

2. Dave Blank 'The High-Tech Revolution' (You Tube / 5 Dec. 2019) *https://youtu.be/ Vs5j0CLPHIl*

3. A transcript can be found online: http://www.phy.pku.edu.cn. The talk was substantially repeated 25 years later, October, 1984, at a seminar entitled 'Tiny Machines Nanotechnology Lecture' (You Tube / 23.Ago. 2012) *https://youtu.be/ 4eRCygdVV--c*

4. IBM 'A Boy And His Atom' (You Tube 30 Apr.2013) *https://youtu.be/oSCX78-8-q0* and *https://youtu.be/xA4QVVwaweVVA*

5. https://www.utwente.nl/en/mesaplus/

The Dark Universe

1. PBS Space Time 'How We Know The Universe is Ancient' (You Tube / 4 may 2020) *https://youtu.be/ Y6Vhh70Lw9w*

2. Adam Rutherford 'A Brief History of Everyone Who Ever Lived' W&N / 2017 p.2

3. 75% hydrogen, 25% helium, plus traces of lithium and a few other bits and pieces.

4. Dan Hooper 'What Happened At The Beginning Of Time' / Royal Institute (You Tube / 5 Mar. 2020) *https://youtu.be/ dB7d89-YHjM*

5. Bill Bryson 'A Short History of Nearly Everything' Black Swan 2004 / p.29-31

6. Andrew Pontzen 'Dark Matter's Not Enough' (You Tube / 19 Nov.2014) *https://youtu.be/GFxPMMkhHuA*

7. Sean Carrol 'The Big Picture' / Royal Institute (You Tube / 22 Nov.2017) *https://youtu.be/2JsKwyRFiYY*

8. PBS Space Time 'How We Know The Universe is Ancient' (You Tube / 4 may 2020) *https://youtu.be/ Y6Vhh70Lw9w*

Bibliography

Recommended Reading

Bill Bryson: 'A Short History of Nearly Everything' / Black Swan 2004

Russel Stannard: 'Relativity A Very Short Introduction' / Oxford University Press revised edition 2017

John Polkinghorn: 'Quantum Theory A Very Short Introduction' / Oxford University Press 2002

Carlo Rovelli: 'Seven Brief Lessons on Physics' / Penguin

Carlo Rovelli: 'Reality Is Not What It Seems' – The Journey to Quantum Gravity / Penguin

Ken Alder: 'The Measure of All Things' / Abacus

Adam Rutherford: 'A Brief History of Everyone Who Ever Lived' / W&N

Andrew Chaikin: 'A Man on the Moon' – The Voyages of the Apollo Astronauts / Penguin 2019 edition

David Barry: 'Sextant: A Voyage Guided by the Stars and the Men Who Mapped the World's Oceans' / William Collins 2015 edition

Dr Richard Gunderman: 'Tesla The Man The Inventor And The Age Of Electricity' / Andre Deutsch 2019

Recommended Video & Podcast

David Tong 'Quantum Fields: The Real Building Blocks of the Universe' / Royal Institute (You Tube / 15 Feb.2017) *https://youtu.be/zNVQfWC_evg*

Andrew Pontzen 'Dark Matter's Not Enough' / Royal Institute (You Tube / 19 Nov.2014) *https://youtu.be/GFxPMMkhHuA*

Sean Carrol 'The Big Picture' / Royal Institute (You Tube / 22 Nov.2017) *https://youtu.be/2JsKwyRFiYY*

Dan Hooper 'What Happened At The Beginning Of Time' / Royal Institute (You Tube / 5 Mar. 2020) *https://youtu.be/dB7d89-YHjM*

Geraint Lewis 'The End of the Universe' / Royal Institute (You Tube / 3 Oct. 2018) *https://youtu.be/IF4UhEIRUFg*

Tara Shears 'Antimatter: Why the anti-world matters' / Royal Institution (You Tube 18 Oct 2013) *https://youtu.be/ 0Fy6oilRwJc*

Dave Blank 'The High-Tech Revolution' / Royal Institute (You Tube / 5 Dec. 2019) *https://youtu.be/ Vs5j0CLPHIl*

Richard Feynman 'The Complete FUN TO IMAGINE with Richard Feynman BBC2' (You Tube / 1 Nov.2018) *https://youtu. be/ P1ww1IXRfTA*

Richard Feynman 'Tiny Machines Nanotechnology Lecture – There´s Plenty of Room at the Bottom' (You Tube / 23 Ago. 2012) *https://youtu.be/ 4eRCygdW--c*

Richard Feynman 'Computer Heuristics Lecture' (You Tube / 3 Jun 2012) *https://youtu.be/EKWGGDXe5MA*

Dario Gil IBM 'The Future of Quantum Computing' (You Tube / 8 May 2020) *https://youtu.be/ zOGNoDO7mcU*

ColdFusion 'Quantum Computers – FULLY Explained!' (You Tube / 27 May 2019) *https://youtu.be/ PzL-oXxNGVM*

CNBC 'The Hype Over Quantum Computers, Explained' (You Tube / 10 Jan. 2020) *https://youtu.be/ u1XXjWr5frE*

Dustin Cavanaugh 'Human Teleportation? Quantum Entanglement' (You Tube 18 Nov. 2016) *https://youtu.be/ hTe2PYwnEpc*

Spark 'How Gravity Affects Time' (You Tube / 4 Jan. 2019) *https://youtu.be/QQRj78jOxWo*

Arvin Ash 'General Relativity Explained simply & visually' (You Tube / 20 Jun. 2020) *https://youtu.be/tzQC3uYL67U*

Arvin Ash 'Visualizing the Planck Length. Why is it the Smallest Length in the Universe' (You Tube / 12 Oct. 2019) *https://youtu. be/ bjVfL8uNkUk*

Fermilab 'Quantum Entanglement: Spooky Action at a Distance' (You Tube /12 Feb. 2020) *https://youtu.be/ JFozGfxmi8A*

PBS Space Time 'How Do You Measure the Size of the Universe?' (You Tube / 25 Feb. 2015) *https://youtu.be/ QXfhGxZFcVE*

PBS Space Time 'How We Know The Universe is Ancient' (You Tube / 4 may 2020) *https://youtu.be/ Y6Vhh70Lw9w*

PBS Space Time 'What are the Strings in String Theory?' (You Tube / 18 Oct. 2018) *https://youtu.be/ k6TWO-ESC6A*

Tech Planet 'Top 7 Emerging Technologies That Will Change Our World!' (You Tube / 17 Apr. 2019) *https://youtu.be/ hTe2PYwnEpcTxRldL2CDBk*

Particle Fever (2013) – CERN (You Tube / 16 Apr. 2019) *https:// youtu.be/ akCJc/K3DUU*

IBM 'A Boy And His Atom' (You Tube 30 Apr.2013) *https:// youtu.be/oSCX78-8-q0* and *https://youtu.be/xA4QWwaweWA*

The Big Bang Theory TBBT S02E02 'Loop Quantum Gravity Vs String Theory' *https://youtu.be/FMSmJCKaaC0*

BBC World podcast series: '13 Minutes to the Moon' / Kevin Fong / www.bbc.co.uk

BBC World podcast series: 'Elements' / www.bbc.co.uk

BBC World podcast series: 'The Bomb' / www.bbc.co.uk

Stephen Fry: 'Great Leap Years' / Podcasts

www.ingramcontent.com/pod-product-compliance
Lightning Source LLC
Chambersburg PA
CBHW022043190326
41520CB00008B/694